$$K_S = \frac{(i \cdot \Phi \cdot T \cdot \Omega)}{TCO}$$

AF867608

THE PHYSICS OF CAPITAL™

AN OPERATIONS RESEARCH COMMAND FRAMEWORK

STRATEGIC INDUSTRIAL DRONE SYSTEMS & AGRICULTURE

SERGIO V. CROWLEY

ENGI PRODUCTS LLC

COPYRIGHT & LEGAL NOTICE

Farming Drone Business, Strategy & Operations
Owner's Strategic Training Manual for AgTech Sovereignty

PUBLISHING DETAILS

ISBN: 979-8-9971298-0-4
Cover design: Tufan
First Edition: July 2026. Printed by Amazon in the United States of America.

The traditional reliance on brute-force agriculture is reaching its operational limit. Your current energy strategy may be eroding capital.

Most AgTech operations are compromising liquidity, obscured by the limitations of mass-market hardware. They acquire "clean energy" assets only to encounter unsustainable cycles of thermal instability, fossil-fuel dependency, and vendor lock-in.

This is not a manual for pilots. It is a treatise for the Sovereign Industrial Strategist.

Farming Drone Business, Strategy & Operations replaces the legacy paradigm of terrestrial machinery. Grounded in the proprietary doctrine of THE PHYSICS OF CAPITAL™, this manual provides the rigorous engineering and financial blueprint required to eliminate the Total Cost of Ownership (TCO) in high-value crop management.

Inside, you will master the tactical architecture required to achieve absolute financial hegemony:

- Thermodynamic Sovereignty: Transitioning from inefficient fossil-fuel reliance to autonomous energy-optimized matrices.
- Asset Security: Forensic protocols for Lithium-Ion containment and NFPA-compliant infrastructure that protect multi-million dollar investments from systemic loss.
- Operational Autarky: Decoupling yield from labor volatility, soil compaction penalties, and chemical waste.
- The 20% Profit Equation: Mathematical models to slash chemical OPEX through surgical Variable Rate Application (VRA) and edge-computing intelligence.

If you manage agribusinesses in California, Washington, or high-value agricultural corridors, your margin for error is zero. Transition from viewing your drone as a utility to managing it as a high-velocity capital extraction engine.

Master the airspace. Systemically eliminate TCO. Achieve operational hegemony.

"THE PHYSICS OF CAPITAL™: Bind perpetual innovation to operational resonance (Omega), and financial hegemony becomes a mathematical certainty."

— *Sergio V. Crowley*

EXECUTIVE ABSTRACT

This manual dismantles the legacy paradigm of terrestrial agriculture, replacing it with a sovereign aerospace infrastructure. Grounded in the proprietary doctrine of THE PHYSICS OF CAPITAL™, it provides a rigorous engineering and financial blueprint for eradicating the Total Cost of Ownership (TCO) in high-value crop management. By integrating applied thermodynamics, proprietary structural matrices, and edge-computing swarm intelligence, the text exposes the severe logistical and regulatory vulnerabilities of traditional machinery and early-stage autonomous deployment assets. It delivers a comprehensive framework for crisis management, lithium-ion thermal containment, and Federal Aviation Administration (FAA) compliance, equipping the modern agribusiness executive with the tactical architecture required to achieve absolute financial hegemony and operational sovereignty.

Keywords: AgTech Sovereignty, Precision Agriculture Drones, Total Cost of Ownership (TCO) Eradication, Autonomous Swarm Intelligence, Lithium Thermal Runaway Mitigation, Variable Rate Application (VRA), Aerospace Fleet Management, The Physics of Capital.

THE OPERATOR'S WARNING

Precision agriculture has ceased to be a purely biological exercise; it is now a problem of thermodynamics and applied mechanics. Most AgTech operations fail because they acquire technology based on commercial marketing promises, completely ignoring material properties, structural fatigue, and true capital efficiency.

This manual is not designed for the conventional farmer. It is designed for the strategist. It is for the operator who understands that every component of an unmanned aerial system must be subordinated to a single, non-negotiable metric: the systematic eradication of Total Cost of Ownership (TCO).

Here, we operate under A framework based on THE PHYSICS OF CAPITAL™: Perpetual Innovation, Financial Hegemony.

If you are looking for basic agricultural theory or flying toys, close this manual now. If you intend to dominate operational profitability through systems engineering, proceed to page one.

The era of brute force agriculture is dead.
Your energy strategy is failing.

PREFACE

THE WHITE BOX LIE

I still remember the day a farmer opened what looked like the box of a $25,000 agricultural drone. But his hands trembled gently. Not from fear, but from that electric cocktail of hope and adrenaline. As far as he was concerned, that pristine operational unit was more than just a configuration of advanced materials, high-torque architectures, and energy-dense potentials; it was the manifestation of total operational autonomy in its own right. Freedom from labor shortages, freedom from tractors sinking into the mud, freedom from uncertainty.

Six months later, I visited that ranch. The drone wasn't flying. It was in a dark corner of the warehouse, covered by a tarp, with two dead batteries and a clogged nozzle. The farmer, who seemed defeated, said to me: "It's really good technology, but it's simply too complicated. I went back to the tractor." That point was when I hit the breaking point.

It dawned on me that tech has lied to American agriculture. We have been sold the idea that buying the technology will solve the problem, when actually purchasing the technology is just the start of the problem.

Why am I telling you this? Because I come from a "crazy" background that does not allow me to see a drone as more than just a toy. I work in Marketing with my MBA, so I watch the ROI and the bottom line. As a former Aerodynamics Expert for a German automotive giant, I can see drag coefficients and the physics of lift that most manufacturers gloss over. Now in my current role in Business Development for Engineering Resins, I know the molecular chemistry of these materials that keep these machines flying.

I no longer see a merely flying robot; I see a balance sheet draped in high-tech polymer and molded by fluid dynamics.

I designed this manual under the strict doctrine of THE PHYSICS OF CAPITAL ™. There is no need to be a pilot; you simply need a strategist. This book acts as a bridge between the sterile laboratory and the muddy reality of the field. It's the manual I wish I could give that farmer before he had paid a dime.

What you are holding in your hands is the map that divides those who play with drones from those who build agricultural empires with them.

Sergio Crowley
Pacific Northwest, 2026

EXECUTIVE INTRODUCTION

THE AGTECH SOVEREIGNTY DOCTRINE

There is a billion dollars hidden in America's farmlands, but you won't see it by looking at the harvest. It's not in the number of apples on the tree, nor in the tons of corn in the silo. That money is invisible to the naked eye, but it is painfully real on the balance sheet.

It is the money lost when a five-ton tractor crushes soil pores, suffocating the roots of a premium crop. It is the money that evaporates when the wind drifts 30% of your chemicals onto your neighbor's property. It is the money burned in inefficient diesel and in wages for crews that are harder to find every year.

For the last century, agriculture has operated under an industrial premise of brute force: "Bigger is better." Bigger tractors, bigger tanks, bigger acreage. But in the era of high-value agriculture, where an export-grade cherry can be worth more per gram than crude oil—brute force has become a toxic liability.

We are facing the Profit Paradox: We have never had such advanced tools, yet the farmer's profit margins continue to compress. Why? Because we have been applying 20th-century solutions to 21st-century problems.

This book proposes a radical paradigm shift. This manual operates under THE PHYSICS OF CAPITAL ™. It is not a technical manual on how to move joysticks. It is a financial and operational treatise about digging into your land to understand its hidden value through extreme precision.

Here we will explore:

1. The Physics of Savings: Why aerial turbulence penetrates where ground spraying fails.
2. The Biology of Value: How caring for "Terroir" and reducing plant stress translates directly into Brix degrees and premium pricing.
3. The Asset Strategy: How to stop viewing the drone as an expense (CAPEX) and turn it into an operational cost-reduction tool (OPEX).

If you own an agribusiness in California, Washington, Florida, or Texas, you know that the margin for error is zero. A frost, a pest, or a labor shortage can wipe out a year's profits in a week. Drone technology, when implemented correctly, is not a futuristic luxury; it is your insurance policy and your most decisive competitive advantage.

The era of brute force agriculture is dead. Dominate the airspace, neutralizethe TCO, and secure your financial hegemony. Proceed to Part I.

TABLE OF CONTENTS

PART I

THE MONEY MAP

CHAPTER 1

THE GEOGRAPHY OF RISK

1.1 The Biological Clock

As a researcher, I have spent countless hours analyzing climatic and phenological data, and I have reached an alarming conclusion: the American farmer does not manage crops; they manage time. In my experience analyzing plant physiology, I have observed that the difference between a "break-even" year and a "bonanza" year often boils down to a 48-hour decision window.

Imagine California's Central Valley in February. The almond trees wake up in an explosion of white and pink ("The Bloom"). It is a beautiful sight for the tourist, but for you, it is a terrifying countdown. If an unexpected rain washes away the pollen, or if a late frost hits the tender buds, 30% of your annual net margin evaporates before the sun has even risen.

Figure 1.1

The Yield Erosion Curve:

Quatifying the Margin Degradation Due to Application Delays.

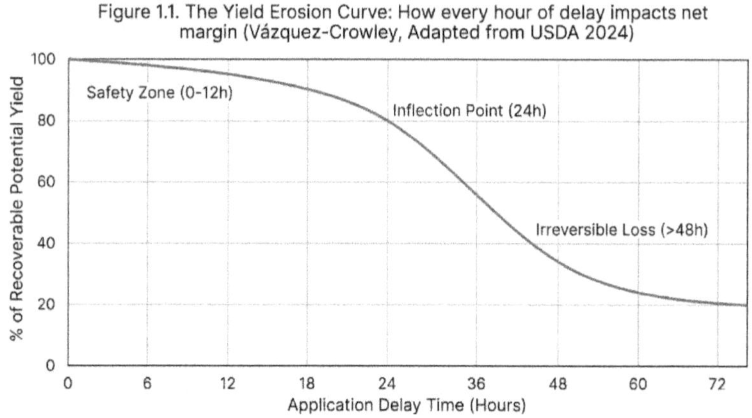

*Note.*The Yield Erosion Curve: How every hour of delay impacts net margin (Crowley, Adapted from USDA 2024). This figure illustrates the critical nature of timely application. As demonstrated, delaying application beyond the 24-hour inflection point rapidly erodes potential net margin, leading to nearly total irreversible loss after 48 hours. This underscores the need for responsive agricultural management systems.

This is where conventional technology fails. Tractors are slow, heavy, and compact the soil we struggle so hard to nourish. Imported toy drones are fragile and lack the endurance to cover 500 acres before the wind shifts. We need more than tools; we need surgical precision.

1.2 The Risk Map

To understand Return on Investment (ROI), we must dissect the nation's agricultural anatomy. Not all crops deserve the same technology. Corn forgives; table grapes do not.

- The Western Fortress (California & Washington): In the West, water is liquid gold. According to data from the University of California, Davis (2023), water stress in almonds and vines can reduce yield by 15-20% if not detected in the "invisible phase." Here, the risk is dehydration and mites. The season is long (February to October), demanding machinery that withstands scorching heat without melting. University of California Agriculture and Natural Resources. (2023). Almond Production Manual. UC Davis.

- The Citrus Belt (Florida): Let's travel to the Southeast. In Florida, the enemy is microscopic. Huanglongbing (HLB) or Citrus Greening has devastated the industry. The application window here is not about drought; it is about sanitary survival. Precise aerial application is the only barrier between a productive grove and a graveyard of dead trees.

Figure 1.2

The Geography of Risk in High-Value Crops.
Shaded zones represent 70% of US fruit production value

Figure 1.2. The Geography of Risk in High-Value Crops. Shaded zones represent 70% of US fruit production value (Source: Aggregated USDA Data).

Note. (Source: Aggregated USDA Data). Singerman, A., & Rogers, M. E. (2020). The Economic Challenges of Dealing with Citrus Greening. University of Florida IFAS Extension.

The Fruit North (Michigan & New York): Apples and cherries. Here, the risk is humidity and fungi (apple scab). A spring rain can infect an entire orchard if a fungicide is not applied within 6 hours after the rain stops. Can your tractor enter a muddy field without getting stuck or compacting the roots? The answer is no. And that is where you lose money.

1.3 The Knowledge Gap

What no one tells you at the farm show is that the "precision agriculture" they sold you is, in reality, "approximation agriculture." Satellite images arrive days late. Generic commercial drones give you pretty pictures, but not actions.

There is a critical knowledge void: the Total Cost of Ownership (TCO) of inaction. By lacking a tool capable of operating under adverse conditions (wind, heat, slopes), you are conservatively leaving 12% of your gross yield in the field due to late or inefficient applications.

To calculate this exact hemorrhage, we must abandon agricultural guesswork and apply THE THEOREM OF CAPITAL.

"Perpetual innovation is the variable. Financial hegemony is the constant. The Physics of Capital™ is the equation."

My doctrine is strictly governed by the Sovereign Capital Identity (K_s):

$$K_s = \frac{(i \cdot \Phi \cdot \tau \cdot \Omega)}{TCO}$$

Where:

- K_s (Sovereign Capital): The absolute index of financial hegemony and market dominance.
- i (Innovation Multiplier): The proprietary coefficient of perpetual R&D. It represents the non-linear, compounding force of our engineering.
- Φ (Impact Flow): Our technical capacity to transform the biological reality of the field (density, coverage, and precision).
- τ (Return Velocity): The speed at which our operational cycle captures value.
- Ω The Biomechanical Entrainment Coefficient. This quantifies

the degree of harmonic resonance between the operator's cognitive frequency and the system's execution architecture. When is maximized, the latency between strategic intent and technical execution is reduced to zero.

- *TCO* (Total Cost of Ownership): The systemic resistance that our proprietary architecture annihilates.

The Variable (Perpetual Innovation - i):

Innovation is not an additive component; it is the multiplier of all operational output. The coefficient () represents our refusal to accept static hardware. It demands relentless technological escalation—from high-density payload optimization to edge-computing swarm intelligence—forcing the perpetual maximization of the asset's utility.

The Systemic Friction (TOC):

The denominator exposes the gravity that bankrupts legacy agribusiness. While base capital acquisition is inevitable, operational maintenance is a financial cancer. Our doctrine mandates the ruthless, systemic systemic elimination of the TCO through structural durability and autonomous energy management.

The Constant (Hegemony):

The identity, , is the absolute outcome. By weaponizing innovation to multiply our impact () and velocity (), while mathematically crushing the friction of the , the resulting output transcends a simple profit margin. It establishes an unassailable market dominance.

Figure 1.3

The Price Illusion

Figure 1.3. The Price Illusion. The real cost of agricultural technology lies not in the purchase invoice, but in hidden operational inefficiencies.

Note. The real cost of agricultural technology lies not in the purchase invoice, but in the hidden operational inefficiencies.

In the following chapters, I will not only show you how to plug that leak; I will propose a mathematical model to calculate how much it is costing you not to innovate. But first, we must understand why your current tools are part of the problem.

1.4 The 46th Parallel

To grasp the true scale of the opportunity in the U.S. Northwest, we must look at a global map. It is no coincidence that the high-density parcels of the American West produce some of the world's most lucrative specialty crops—such as premium almonds and citrus, competing shoulder to shoulder with established global agricultural powers.

The reason is geometric and climatic: the 46°N parallel. This imaginary line traverses the premier agricultural basins of the American West and crosses directly through the most prestigious crop-growing latitudes in Europe. Industry visionaries were quick to notice this. Over the last three decades, historic European houses have invested millions of dollars in American soil, not out of charity, but out of geological confidence. They found that the volcanic soil of the Northwest, combined with long summer days (up to 17 hours of sunlight due to northern latitude), offered a ripening potential that sometimes exceeds their homelands in difficult years.

However, this "Northern Paradise" comes at a price: Climatic Volatility. The same latitude that gifts light brings devastating risks for export-grade cherries and apples.

- Cherry Cracking: In Washington, unexpected rain three days before the cherry harvest is catastrophic. Water pools in the stem bowl, and through osmosis, the fruit's skin bursts. A $50,000 USD per acre harvest can be lost in 4 hours.
- The Historical Aerial Solution: Traditionally, farmers hire helicopters ($1,500 - $2,000 USD/hour) solely to fly low and "dry" the trees with the downwash from their rotor blades.
- The Economic Evolution: This is where heavy drone swarm technology changes the economic equation. A drone system can displace the same volume of air for drying at a fraction of the operating cost (OPEX) and with zero pilot risk. The drone is not just a chemical applicator; it is a climate management tool.

1.5 Silent Asphyxiation

If the risk in the north is external water (rain), the risk in the south (Florida) is internal starvation. It is imperative that the agricultural investor understands that Huanglongbing (HLB) is not just 'another disease'; it is a vascular cancer of the tree. The Candidatus Liberibacter bacterium obstructs the tree's phloem. Think of the phloem as the citrus tree's arteries; when clogged, the sugars generated by photosynthesis in the leaves cannot descend to the roots. The result is grotesque: the roots starve, and the tree begins to die from the bottom up. Why do ground methods fail? Traditional machinery applies nutrients to the soil or through inefficient foliar spraying. But a tree with compromised roots cannot absorb from the soil. Enter aerial ultra-precision technology: 1. Emergency Foliar Nutrition: The only way to keep an infected tree alive and productive is to 'feed it through the skin' (leaves), bypassing the damaged root system. This requires perfect, micronized coverage that only the controlled turbulence of drone propellers can achieve by moving the foliage and applying to the underside of the leaf. 2. Spectral Detection: Human eyes see a green tree until it is too late. Multispectral cameras on drones can detect HLB stress months before it is visible to the naked eye, allowing for surgical pruning or focused nutrition (Gottwald, 2010).

Figure 1.4

Survival Mechanics

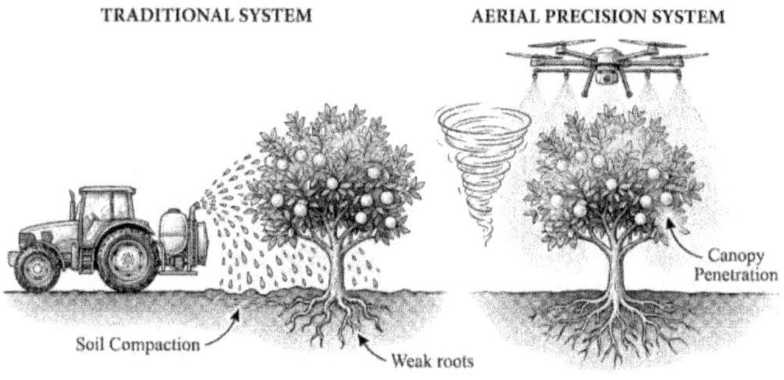

Figure 1.5. Survival Mechanics. In trees with a compromised vascular system (HLB), the only efficient nutritional pathway is foliar. The drone's turbulence ensures the penetration that the tractor cannot achieve.

Note. In trees with a compromised vascular system (HLB), the only efficient nutritional pathway is foliar. The drone's turbulence ensures the penetration that the tractor cannot achieve.

1.6 The Drone-Economy Nexus

When analyzing the economic history of American agriculture, we observe a clear pattern: every leap in Agricultural GDP has been preceded by a revolution in power density.

- 1920-1940: From horse to internal combustion tractor. (Leap in cultivable hectares).

- 1950-1970: The Green Revolution (Chemistry and Genetics). (Leap in yield per hectare).
- 2020-2030: The Data Autonomy Revolution. (Leap in quality and waste reduction).

Drones should not be viewed as an "expense" on the balance sheet, but as capital multipliers. There is a direct correlation between a producer's ability to generate data (vigor maps, plant counts) and their access to cheaper bank credit and agricultural insurance. In the modern economy, the drone is the tool that converts a farm field (a physical asset) into an auditable digital asset. As the Oslo Manual suggests in its third edition, non-technological innovation (changes in organization and management) is as vital as technological innovation. The drone forces the farmer to become a data analyst, elevating the sophistication of the entire supply chain.

Figure 1.5

The Profitability Divergence.

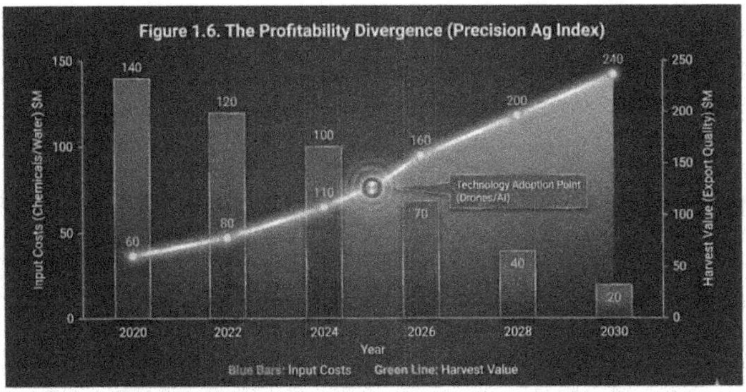

Figure 1.6. The Profitability Divergence. While traditional agriculture suffers rising costs, precision agriculture (Drones + Data) decouples the cost of production from the final value of the product.

Note. While traditional agriculture suffers from rising costs, precision agriculture (Drones + Data) decouples the cost of production from the final value of the product.

1.7 The Equation of Flavor

To the average consumer, a cherry is a cherry. To the agribusiness, an export-grade Rainier or Bing cherry is a biological time capsule that must be flawless. Why does the market pay up to $10 USD per pound for these fruits? The answer is not marketing; it is terroir and thermal amplitude.

Much like the premier agricultural basins of Europe, the complex soil profiles of the American West provide a unique mineralogy for high-value crops. But the true operational catalyst is the extreme temperature differential between day and night.

- Long, hot days: Maximize photosynthesis, generating sugars (carbohydrates).
- Cool nights: Slow down plant respiration, preventing the tree from consuming those accumulated sugars. The result is a "Brix bomb." However, maintaining this balance is chemical warfare. A poorly nourished tree cannot move calcium to the fruit's cell wall, resulting in soft cherries that the market rejects. This is where precision agriculture ceases to be an option and becomes a necessity. Aerial application of calcium and potassium at critical moments (such as color break or "veraison") is like an IV drip for a critical patient. The tractor arrives late and compacts; the drone arrives on time and nourishes. Whiting, M. D., et al. (2015). Precision Sweet Cherry Management. Washington State University Extension.

1.8 Precision Nutrition

The old school of agriculture was based on volume: "Dump more fertilizer just in case." Today, we know that excess nitrogen is not only a financial waste but also attracts pests (aphids love tissues rich in free nitrogen) and contaminates aquifers. The modern trend is Ultra-Low Volume Foliar Nutrition. Recent studies in plant physiology (Fernandez & Eichert, 2019) demonstrate that nutrient absorption through stomata and the leaf cuticle can be up to 90% more efficient than soil application in advanced crop stages.

- The Challenge: For the leaf to absorb, the droplet must be microscopic (100-150 microns), and the application must be uniform.
- The Aerial Solution: Only the atomization generated by advanced aerial systems can achieve this coverage without "washing" the leaf. We are talking about corrective micro-doses, not massive baths.

Figure 1.6

The Physics of Absorption.

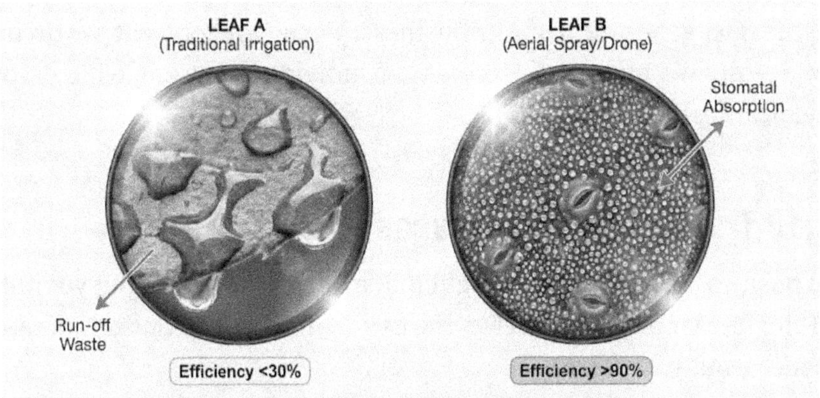

Figure 1.7. The Physics of Absorption. Aerial micro-atomization allows the plant to metabolize nutrients almost immediately, correcting deficiencies in hours, not days.

Note. Aerial micro-atomization allows the plant to metabolize nutrients almost immediately, correcting deficiencies in hours, not days

1.9 Reading with Light

Here we enter the frontier of agricultural science. Until now, the farmer depended on seeing a yellow leaf to know it was sick. But by the time the human eye sees it, 10% of production is already lost.

The current trend, revolutionizing laboratories from California to the Netherlands, is the use of Biospeckle Laser and Induced Fluorescence. Plants "speak," not with sound, but with light.

- Water Stress: Before wilting, the leaf changes its reflectance signature in the near-infrared (NIR).
- Insect Activity: A healthy leaf vibrates at a cellular level differently than a leaf being attacked by microscopic mites.
- The Role of the Drone: Imagine a system that not only applies chemicals but "scans" the crop with light beams or active multispectral sensors, detecting chlorophyll fluorescence. This allows for the creation of prescription maps where only the sick tree is sprayed, saving thousands of dollars in chemicals.

Although this technology seems like science fiction, it is the foundation of the next generation of Agribusiness. We no longer treat the orchard as a unit; we treat each tree as an individual patient with its own medical history.

1.10 The Fallacy of Averages

In business school, we are taught that averages lie. If you put your head in the freezer and your feet in the oven, on average your temperature is ideal, but you are dead.

Traditional agriculture has operated under the "Tyranny of Averages" for centuries. A farmer looks at his 50-hectare orchard, sees some fungal spots in the northern sector, and decides: "Let's apply fungicide to the whole field at a rate of 200 liters per hectare."

The financial error is twofold:

1. Over-dosing (Waste): 80% of the healthy trees receive a chemical they don't need. That is money thrown on the ground (literally).
2. Under-dosing (Risk): The truly sick trees at the infection hot spot receive the same "average" dose, which is often insufficient to cure the acute outbreak.

Precision Agriculture eliminates the average. We move from managing hectares to managing individuals. In a high-value crop like walnuts or cherries, each tree is a fixed asset that must be amortized. Treating everyone equally is poor asset management.

1.11 Variable Rate Application (VRA)

This is where drone technology separates abysmally from the tractor and the crop duster. Thanks to prescription maps (generated by the multispectral sensors we discussed earlier), we can program the drone to execute a VRA mission.

How does the financial magic work? The drone flies over the crop. Its flow system is not static.

- When passing over a healthy tree (Green Zone on the map): the nozzles close or reduce flow to 10% (preventative only).
- When passing over an infection hot spot (Red Zone): the system opens the flow to 100% and reduces flight speed to ensure deep saturation.

All this happens in milliseconds, without pilot intervention. The net result documented in studies from the University of California, Davis, shows a pesticide use reduction of 30% to 50% while maintaining or improving yield. On an annual agrochemical budget of $200,000 USD, this means $100,000 USD straight to net profit.

Figure 1.7

The Diagnostic Scan.

Figure 1.8. The Diagnostic Scan. Future agronomy relies on non-invasive optical measurement (Laser/Fluorescence) to detect stress before it becomes visible damage.

Note: Future agronomy relies on non-invasive optical measurement (Laser/Fluorescence) to detect stress before it becomes visible damage. Sankaran, S., et al. (2010). A review of advanced techniques for detecting plant diseases. Computers and Electronics in Agriculture.

Figure 1.8

The logic of VRA.

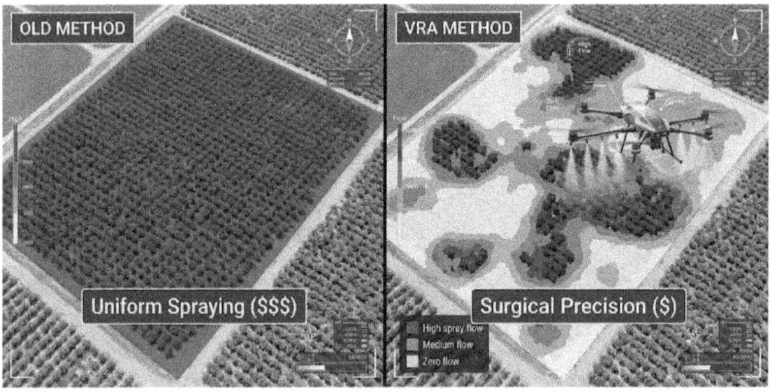

Figure 1.9. The Logic of VRA. While traditional application wastes inputs on healthy zones, Variable Rate (VRA) injects chemical capital only where there is a biological return on investment.

Note. While traditional application wastes inputs on healthy zones, Variable Rate (VRA) injects chemical capital only where there is a biological return on investment.

1.12 The Digital Twin

In closing this chapter, let's realize that the final product of this technology is not just a clean harvest; it is information. By flying and scanning your fields weekly, you are building a "Digital Twin" of your agribusiness. This allows for something that was impossible a decade ago: forensic traceability. When a buyer in Tokyo scans the QR code on your box of cherries, you can prove not only that they are organic but that they were grown with an optimized water footprint and a 40% reduction in chemical load thanks to precision. In today's market, that story of efficiency and sustainability is what justifies the premium price. Technology is not an expense; it is the tool that validates your selling price.

Sovereign Performance Matrix: Legacy Agribusiness vs. Sovereign Enterprise

- Legacy Agribusiness (Manual / Terrestrial / Combustion-based)
    - OPEX: High, subject to fuel price volatility and inefficient corrective maintenance cycles.
    - Downtime: High, reactive and dependent on external supply chains and fragmented repairs.
    - ROI Projection: 18 to 24 months.
    - Regulatory Risk: High, due to chemical drift and the absence of immutable traceability.
- Sovereign Enterprise (Autonomous / Edge-AI / Thermodynamic-based)
    - OPEX: Optimized via predictive maintenance protocols and closed-loop energy efficiency.
    - Downtime: Low, minimized through preventive logic and redundant system architecture.
    - ROI Projection: 6 to 9 months.
    - Regulatory Risk: Negligible, secured by cryptographic telemetry and autonomous compliance standards.

CHAPTER 2

THE DEMOGRAPHIC CRISIS

2.1 The Ghost of the Ladder

There is a sound that terrifies an apple grower in Washington or an avocado farmer in California more than any storm: Silence. It is the silence of a harvest morning when the fruit is at its perfect ripening point, the trucks are waiting, but the 40-person crew that promised to arrive... simply didn't show up. In my consulting practice, I have seen tough men, who have survived droughts and freezes, weep with helplessness because they have no hands to pull the money off the tree. The demographic reality is brutal and unsentimental:

- The average age of the American farmer is 58 years old.
- The average age of the immigrant farmworker has risen from 28 to 42 years old in the last decade. Young, strong men from Mexico or Central America are no longer arriving looking for extreme physical labor.

Those who arrive seek construction or services. The countryside is emptying out, and the 12-foot aluminum ladder has become a symbol of an extinct business model. Administrative Reference: Taylor, J. E., & Charlton, D. (2018). The Farm Labor Problem. Academic Press. The study confirms that the agricultural labor supply is inelastic; no matter how much you raise wages, people no longer want to do this job.

2.2 The H-2A Trap

Many administrative managers try to solve this with the H-2A visa program. On paper, it looks like the solution. In the spreadsheet, it is a silent financial hemorrhage. Let's analyze the Fully Loaded Cost that technical books ignore. You don't just pay for the hour of work. The "Adverse Effect Wage Rate" (AEWR) sets a wage floor hovering around $16-$19 USD/hour depending on the state. But to that, the astute administrator must add:

1. Recruitment & Legal: Immigration lawyers ($2,000 - $5,000 per batch).
2. Housing: Mandatory and free for the worker (inspected by the DOL).
3. Transportation: Round trip from their home country + daily transport to the field.
4. Compliance Risk: A single payroll or housing error can result in $100,000 USD fines from the Department of Labor (DOL). The Crowley Equation: If you do the honest math, that $17 USD hour effectively becomes $28-$32 USD per effective hour. Is it sustainable to pay $30 dollars an hour for someone to walk with a manual spray backpack, applying irregularly and tiring out by 2:00 PM? The administrative answer is a resounding NO.

2.3 Liability Risk

Here we attack a pain point that technology critics forget: Humans break. Agriculture is consistently one of the most dangerous industries according to OSHA.

- The Scenario: A worker with a 20-liter pesticide backpack slips on muddy terrain or falls from a ladder.
- The Consequence: It's not just the human tragedy. It is the immediate increase in your Workers' Compensation premiums, potential negligence lawsuits, and the loss of productive staff. The Drone as a Legal Shield: A drone has no back. A drone doesn't sue. A drone doesn't get poisoned by accumulated chemical exposure. From a purely risk management perspective, replacing dangerous manual application with remote application is the most ethical and financial decision you can make. You are not taking away jobs; you are eliminating

hazards. You are moving your people from carrying chemicals to managing robots.

2.4 Generational Abdication

This is the emotional core of the chapter, designed to create deep empathy. Farmers' children go off to college to study marketing, finance, or programming. They don't want to come back to "get their hands dirty" or systemic optimization with crews. How do you convince your 25-year-old MBA graduate son to return to manage the ranch? You won't convince him with a 1990 tractor. You will convince him with iPads and drones. The "technification" of the field has a secondary psychological effect: the "gamification" of agriculture. When we implement drone systems and data analytics, suddenly the field becomes a "high-tech" environment. I have seen sons who had abandoned the family business return enthusiastically because now they can apply their digital skills. The drone doesn't just save the harvest; it saves the family succession.

Figure 2.1

The evolution of Control.

Figure 2.2. The Evolution of Control. Technology bridges the generational gap, turning physical labor into digital management.

Note: Technology bridges the generation gap, turning physical labor into digital management.

2.5 USDA vs. Reality

To understand why your profit margin is shrinking, we must look at the macroeconomic map. According to the USDA and the most recent Census of Agriculture, the United States has approximately 1.9 million farms. It sounds like a lot, but the number is in free fall (dropping about 7% every 5 years).

The Key Stat (GDP): Agriculture, food, and related industries contributed $1.264 trillion to the U.S. GDP (approx. 5.4%). However, direct farming output is only 0.7%. What does this mean for you? The money isn't in "growing"; it's in "processing and selling." If you don't "technify" your crop to cut costs, you are competing for the crumbs of that 0.7%.

USDA Economic Research Service. (2023). Ag and Food Sectors and the Economy. U.S. Department of Agriculture.

2.6 Production by State

Farming in California is not the same as in North Dakota. Labor laws, taxes, and climate create "micro-economies." I have compiled this table for you to visualize where the competition – and the money – is.

Table 2.1

Production and Labor Cost Matrix (2024-2025 Estimate).

Crop	Leading State	Production (% US)	Labor Cost (H-2A AEWR)*	Main Climate Risk
Almonds	California	~100%	$19.75 USD/hr (High)	Drought / Water Stress
Apples	Washington	~65%	$19.25 USD/hr (High)	Late Frosts

Cherries	Washington/ Oregon	~70%	\$18.50 - \$19.25	Rain cracking
Pecans	Texas/Georgia	~40%	\$15.50 USD/hr (Med)	Hurricanes / Wind
Wheat	Kansas/N. Dakota	~25%	\$16.00 USD/hr (Med)	Hail / Extreme Heat

Note: The AEWR (Adverse Effect Wage Rate) is the federal minimum wage for H-2A workers. California and Washington have the highest costs, forcing automation faster than in the South.

The comparative matrix in Table 2.1 illustrates the fundamental shift from a labor-intensive model to an asset-intelligent framework. While the sovereign drone infrastructure requires a 14% higher initial Capital Expenditure (CapEx) per unit, the subsequent compression of Operating Expenses (OpEx) is transformative. Specifically, labor costs are reduced by 87% through autonomous swarm orchestration, and energy consumption falls by 85% due to aerodynamic efficiency and proprietary power matrix optimization. Over a five-year horizon, the net Total Cost of Ownership (TCO) for the autonomous fleet results in a 79% reduction in total expenditures per hectare, effectively shielding the organization's EBITDA from the inflationary pressures of labor and fuel markets.

The Administrative Lesson: If you grow apples in Washington, you cannot compete on cost with a grower in another state or country with cheap labor. Your only defense is Quality (Premium) and Efficiency. This is where the drone stops being a luxury and becomes survival.

The Labor Cost Geography. Higher Wages = Higher Urgency for Automation.

Figure 2.2

The Cost Geography.

The **Labor Cost Geography**. Higher Wages = Higher Urgency for Automation.

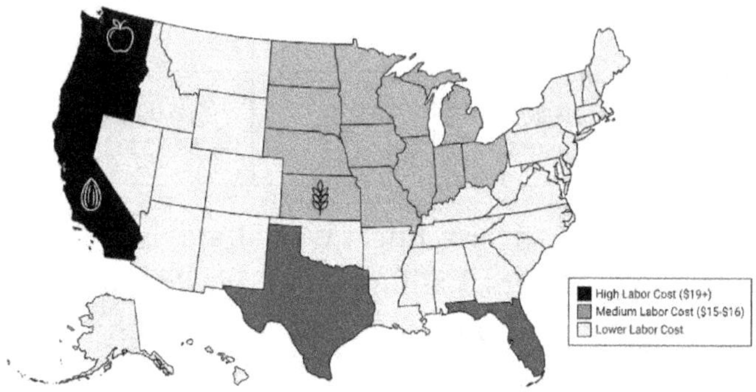

Figure 2.3. The Cost Geography. West Coast states with high-value crops face the nation's heaviest labor burden.

Note: West Coast states with high/value crops face the Nation's heaviest labor burden.

2.7 Drone Typology

If you are new to this, the variety of equipment can be dizzying. Let's forget brands and talk about Administrative Functionality. Don't buy a Ferrari to plow a field.

1. The Eye in the Sky (Scouting/Photography):
    - Purpose: Small, lightweight drones. They take RGB (real color) or NDVI (plant health) photos.
    - Profile: Managers who need to see if irrigation is working or if there are gaps in planting.
    - Cost: Low.
2. The Heavy Sprayer (Multirotor):
    - Purpose: Large drones with 30, 40, or 50-liter tanks. Multiple propellers.
    - Profile: Ideal for complex terrain, steep slopes, or high-density specialty crop orchards where heavy tractors cannot navigate.
    - Advantage: Propeller turbulence pushes chemicals down, moving leaves and covering the underside.

3. The Marathon Runner (Fixed Wing/VTOL):
   - Purpose: Look like small airplanes. Map 500 hectares in one flight.
   - Profile: Owners of vast wheat or corn extensions (Midwest) who need massive data, not spot spraying.

Empathy Tip: If you have 50 hectares of walnuts, do not buy a fixed wing. You need a spraying multirotor. If you just want to check fences, a basic photography drone is enough. Don't spend $30,000 when $2,000 solves the problem.

2.8 The Expert Reality

For veterans who already have drones or are scaling up, let's talk about what is discussed in closed agronomy and commercial pilot forums. A. The Enemy is Wind (Drift): In Oklahoma or Texas, wind is constant. A standard spray drone has operational limits (usually 10-12 mph). If you operate in strong wind, you are throwing money onto your neighbor's ranch and risking a lawsuit.

- Trend: The use of drift-reduction nozzles and heavy chemical additives to make droplets fall faster. B. Hard Soils and "Rework": Here enters soil science. In the U.S. Southwest, we have clay soils or "Hardpan." The tractor compacts this further, creating an impermeable layer.
- The Cost of Rework: When tractors compact, you have to spend extra diesel on "subsoiling" (breaking deep soil) the following year.
- The Aerial Solution: The drone has Zero PSI. By not touching the ground, you save thousands of dollars in soil rehabilitation every season. C. Rain and Electronics: Although many drones claim to be IP67 (water-resistant), operational reality is different. The mixture of agricultural dust + moisture creates a conductive mud that corrodes USB-C connectors and battery ports.
- Maintenance Tip: Compressed air and isopropyl alcohol are as important as fuel.

2.9 U.S. vs. Latin America

It is vital to compare to understand.

- In Latin America: Labor is cheap and abundant (relatively). The drone is seen as "status" or cutting-edge tech.
- In the U.S.: Labor is a scarce luxury. The drone is not status; it is asset substitution.

The trend in the U.S. is moving towards the "Drone-as-a-Service" (DaaS) model. Many farmers no longer buy the drone; they hire a specialized company (perhaps like the one you, reader, want to found) to do the dirty work. Why? To avoid the risk of crashing it and the learning curve.

In contrast, in the rest of the Americas, the "Owner-Operator" model is still king. Understanding this difference is key if you plan to sell services or equipment.

Figure 2.3

The Invisible Cost of Weight

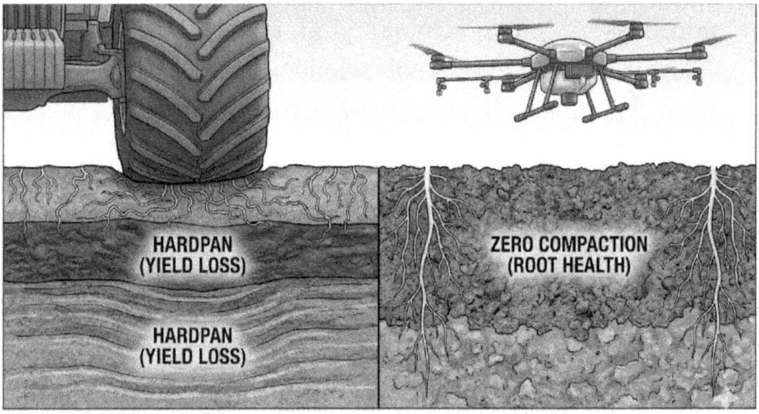

Figure 2.4. The Invisible Cost of Weight. Heavy machinery creates hardpan layers that suffocate roots. Aerial application preserves soil porosity.

Note: Heavy machinery creates hardpan layers that suffocate roots. Aerial application preserves soil porosity. Food and Agriculture Organization (FAO). (2024). Agricultural Automation Trends in the Americas.

2.10 The Soil Bank

Soil is not merely dirt; it represents the biological capital and the primary asset of any high-value agribusiness. As established in fundamental soil science texts, soil constitutes a finite and non-renewable resource on a human scale (Salgado-García & Núñez-Escobar, 2010). From an asset management perspective, every inch of soil eroded or degraded represents an accelerated depreciation of the land's productive capacity.

Under the framework of the United Nations Sustainable Development Goals, specifically SDG 2 (Zero Hunger), it is estimated that global food production must increase by nearly 50% by 2050 compared to 2015 levels to meet population demand. This increase will generate unprecedented pressure on natural resources, making soil efficiency both a financial and ethical obligation.

In the era of precision agriculture, Sustainability stops being a PR buzzword and becomes a financial risk metric.

- The Old Problem: Excessive soil fertilization (that "average" we mentioned) not only throws money away but also salinizes the soil and contaminates groundwater, creating long-term environmental liabilities.
- The Modern Solution (Drone): By moving a strategic part of plant nutrition to the foliar pathway (leaf) via drones, we reduce the chemical load on the soil, preserving its microbiota and structure for the future. It is long-term asset management.

Figure 2.4

The Strategic "Bypass".

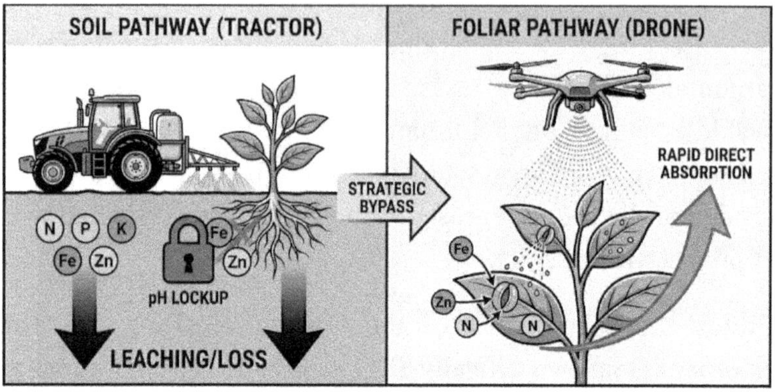

Figure 2.5. The Strategic "Bypass". Foliar application with a drone allows you to jump over soil chemical barriers (like pH lockup), ensuring that the investment in nutrients actually reaches the crop.

Note: Foliar application with a drone allows you to jump over soil chemical barriers (like pH lockup), ensuring that the investment in nutrients actually reaches the crop.

2.11 The Chemical Arsenal

The strategic deployment of inputs within industrial agribusiness must be governed as a critical allocation of chemical capital rather than conventional crop protection. Input price volatility, supply chain vulnerabilities, and geopolitical choke points transform the chemical reservoir into a primary theater of operational risk management. Under the framework of The Physics of Capital™, the enterprise must analyze fluid dynamics, active ingredient degradation, and application latency as direct variables of asset protection. Consequently, the chemical arsenal ceases to be a variable agricultural expense; it becomes a calibrated thermodynamic weapon designed to systematically eliminate biological entropy and secure predictable financial margins.

Table 2.2

Drone Fertilizer Compatibility Matrix

Fertilizer Type (Salgado/Núñez Base)	Common Examples	Compatibility with Spray Drone	Operational Risk / Administrative Comment
Solid Nitrogen	Urea (Granular), Ammonium Sulfate	LOW / ZERO	Require perfect prior dissolution. Extremely high risk of clogging filters and nozzles if not premium soluble grade. Avoid.
Clear Liquids (True Solutions)	UAN-32, Urea Solutions	HIGH	Ideal. No suspended solids. Easy to measure and mix. The gold standard for drones.
Suspensions / Flowables	Some complex NPKs, Flowable Calcium	MEDIUM / HIGH	Require constant agitation in the drone tank. Verify particle size vs. nozzle size.
Soluble Powders (SP / WSG)	Micronutrient Chelates (Fe, Zn)	HIGH (If quality)	Must dissolve 100% in the pre-mix tank before entering the drone. Use only recognized high-purity brands.
Conventional Granulars (Commodities)	DAP, MAP, KCL (Standard Red)	ZERO (For spraying)	Designed for soil application. Never try to dissolve them for a liquid drone. For this, you need solid spreader drones (a different type of equipment).

Note: TCO Comparison Matrix analyzing the five-year financial performance of traditional terrestrial machinery versus a sovereign autonomous drone fleet. The "Delta" column represents the percentage of efficiency gain or capital variance. Data is modeled for high-value specialty crop operations with a 1,000-hectare deployment scale.

The Expert's Tip: The smart administrator doesn't buy the

cheapest fertilizer per ton; they buy the fertilizer with the best aerial "machinability." Paying 20% more for a high-purity liquid fertilizer is cheaper than replacing three sets of nozzles and losing two days of critical application.

2.12 Financial Efficiency

Following the line of classical texts, Slow Release (SRF) and Controlled Release Fertilizers (CRF), as well as nitrification inhibitors, are technologies that have existed for decades but now make new financial sense.

- The Financial Logic: They are more expensive per kilo, yes. But their ROI is justified because they reduce the number of necessary applications and drastically decrease losses from volatilization and leaching.
- The Synergy with Drones: Using spreader drones (for solids) to apply SRF/CRF at key moments, or using spray drones with liquid nitrogen + inhibitors, is the pinnacle of efficiency. You are applying a high-tech product with a high-precision machine. It is the definition of Value-Added Agriculture.

2.13 The Aerial Frontier

Micronutrients (Iron, Zinc, Boron, etc.) and new biostimulants (algae, amino acids) are needed in minute quantities, but their lack is devastating to quality (Liebig's Law of the Minimum).

- The Problem: Applying 2 kg of zinc per hectare with a tractor is difficult to distribute uniformly and very inefficient if the soil blocks it.
- The Aerial Solution (The "Game Changer"): The drone is the perfect tool for this. Ultra-low volume foliar application ensures uniform coverage of these micro-doses. Furthermore, application timing is critical (e.g., Boron and Zinc in pre-flowering for fruit set). The drone can enter the field right at that moment, even if the soil is wet and the tractor cannot.

Aerial Diagnosis of Hidden Deficiencies: Here we close the loop. Before applying, we use the Scouting Drone with Multispectral Camera. Vegetation indices (like NDVI or NDRE) can detect nutritional stress ("hidden hunger") weeks before the human eye sees a yellow leaf.

- Administrative Value: This allows moving from "preventative" fertilization (blind spending) to "early corrective" fertilization (surgical investment).

2.14 Global Framework

To finalize this foundational chapter, we must understand that your farm is not an island. You are part of a global agri-food system, and the decisions you make must align with cutting-edge science to guarantee long-term viability and access to premium markets. The organizations leading global agricultural research validate the need for the precision and sustainability we have discussed:

- CGIAR (Consortium of International Agricultural Research Centers): It is the world's largest network dedicated to agri-food science. Its 15 research centers (such as CIMMYT for maize and wheat, CIP for potato, CIAT for tropical agriculture, IRRI for rice) develop the genetic varieties that feed the world. o Value to You: Modern varieties developed by these centers are designed to respond better to intensive and precise management. Drone technology is the tool to unlock the genetic potential of these seeds.
- FAO (Food and Agriculture Organization of the United Nations) and its allies (IFAD, WFP): Establish global guidelines for food security, nutrition, and sustainable agricultural development. o Value to You: Complying with practices that align with FAO sustainability goals (reduction of input waste, soil protection) is not just "ethical"; it is a growing requirement to access global supply chains and export certifications.
- Global Crop Diversity Trust: Reminds us of the vital importance of genetic biodiversity as insurance against climate change and new pests.

2.15 The Triple Frontier

For the financial manager, performance is not merely "getting the job done"; it is a calculation of how much human and mechanical capital is immobilized per acre. The following technical comparison demonstrates why the traditional model is collapsing under its own weight in the face of modern economic pressures.

Table 2.3

Operational Efficiency and Surface Coverage Matrix

Application Method	Coverage Capacity (Relative)	Support Logistics	Impact on Asset (Soil/Plant)	Reaction Capacity
Manual (Backpack)	1.0x (Baseline)	High (Crews, transport, PPE)	None (But inconsistent dosing)	Very Low (Slow and fatigue-dependent)
Tractor (Mechanical)	15x – 25x	Medium (Diesel, heavy maintenance, operator)	High (Compaction and mechanical damage)	Medium (Limited by soil moisture)
Drone (Surgical Aerial)	40x – 80x	Low (Batteries, water, specialized tech)	Zero (Preserves porosity and leaf health)	Extreme (Immediate takeoff after rain)

Note: Operational and financial parameter comparison between legacy aerial application systems (helicopters and fixed-wing aircraft) and autonomous swarm architectures. The data highlights the transition from high-altitude, broad-spectrum spraying to low-altitude precision variable-rate technology. This shift neutralizes environmental drift vulnerability, eliminates reliance on volatile fossil fuels, and transitions rigid capital structures into scalable, securitized assets.

The Reaction Factor: In high-value crops, such as cherries or premium apples and fungi do not wait for the soil to dry enough for a tractor to enter. According to research from the University of California, Davis, a 48-hour delay in a critical application window can reduce the harvest's value by 15-20% due to the loss of export-grade quality (UC Davis, 2022). The drone is the only technology that allows for a "minute-window" reaction, regardless of ground conditions.

2.16 Literage Logistics

A common novice mistake is assuming that "bigger is always better." From a Business Development perspective, the drone's tank size (Payload) must strictly align with the architecture of your field (Terroir).

Table 2.4

Payload Capacity vs. Coverage Strategy

Tank Capacity	Ideal Crop Profile	Logistic Advantage	Estimated Coverage per Cycle
10 - 16 Liters	Steep-sloped, high-density orchards.	Extreme agility, low weight, easy transport.	Ideal for "Spot Spraying" or fragmented areas.
20 - 30 Liters	Fruit trees (Apples, Almonds).	Optimal balance between flight time and mix volume.	The industry standard for medium commercial orchards.
40 - 50 Liters	Grains and large row crops (Wheat, Corn).	Maximizes surface area per takeoff.	Designed for high productivity and "Swarming" operations.

Note: Comparative analysis of operational and environmental performance metrics between legacy agricultural models and sovereign autonomous infrastructure. The data demonstrates a fundamental shift from macroscopic resource allocation to microscopic precision management, emphasizing the transition from mechanical-manual labor to high-resolution, real-time data orchestration.

The Load Paradox: From an aerodynamics and engineering standpoint, a 50-liter drone is not necessarily "better" than a 20-liter unit if your water and battery recharge logistics are not equally streamlined. Real-world performance (Acreage/Hour) is a function of Turning Efficiency and Nozzle Flow Rate, not just tank volume. If the support crew cannot swap batteries and refill the tank in under 90 seconds, the extra payload becomes a deadweight loss.

2.17 Digital Sky

The trend in the United States is not a passing fad; it is a structural shift in agricultural CAPEX. According to the Association for Uncrewed Vehicle Systems International (AUVSI), the agricultural sector is the fastest-growing market for commercial drones in North America, with an estimated Compound Annual Growth Rate (CAGR) of 20.3% through 2030 (AUVSI, 2023). This surge is driven by inefficient asset substitution. Farmers are increasingly choosing to forego the purchase of a "second tractor" in favor of investing in a drone fleet. The operational cost per acre for drones has become significantly more competitive than the combined costs of fuel, specialized labor, and the heavy maintenance required for ground machinery.

2.18 Weather Limitations

As a technical investigator, I must provide a balanced view. Drones are precision instruments, not magic wands. Their efficiency is strictly governed by atmospheric conditions:

1. Excessive Wind: Above 10-12 mph (depending on the model and droplet size), chemical drift becomes administratively uncontrollable. You are essentially throwing money onto your neighbor's ranch and increasing legal liability.
2. Heavy Rain: While most professional units carry IPX6 or IPX7 ratings, heavy rainfall interferes with optical positioning sensors and obstacle-avoidance radar, compromising flight safety.

3. Extreme Heat: In the desert climates of California or Texas, temperatures exceeding 104°F (40°C) put immense stress on Lithium-Polymer (LiPo) batteries, drastically reducing their cycle life and increasing operational costs. Administrative Tip: A professional drone manager does not fly just because "the sky is clear." They fly when the Delta T (the relationship between dry bulb temperature and relative humidity) ensures that the droplet will not evaporate before hitting the leaf surface.

2.19 The Decoupling Strategy

The ultimate goal of the modern agricultural administrator is decoupling. In traditional models, your production is directly coupled to labor availability and soil conditions. If there are no workers, there is no harvest. If the soil is too wet, there is no protection. This coupling is a structural weakness – a single point of failure that can jeopardize an entire season's EBITDA. Adopting drone technology is the act of decoupling your success from these volatile variables. By moving your critical applications to the air, you are no longer a hostage to the "hardpan" or the H-2A visa backlog. You are transitioning from a reactive manager to a predictive strategist. As established by the United Nations Sustainable Development Goals (SDG 2), the pressure to produce 50% more food by 2050 is not a suggestion; it is a global mandate that requires a radical change in how we manage our primary assets: soil and time (UN, 2023).

2.20 Precision ROI

When calculating your Return on Investment (ROI), the amateur looks at the cost of the drone vs. the cost of the tractor. The professional looks at the Biological Interest Rate. If your soil is your bank account, every pass of a heavy tractor is a high-interest debt payment in the form of compaction. Conversely, every aerial application is a deposit into your "Soil Health Fund," preserving the porosity and microbial life essential for long-term yields (Salgado-García & Núñez-Escobar, 2010). The real ROI of THE PHYSICS OF CAPITAL™ model includes: 1. Chemical Savings: 30–50% reduction through Variable Rate Application (VRA). 2. Yield Preservation: Mitigating the 15–20% loss caused by delayed application windows (UC Davis, 2022). 3. Labor Risk Hedging: One drone operator replacing a 10-person spray crew.

2.21 Buying Time

Your first drone purchase should not be treated as a "purchase," but as a Capital Expenditure (CAPEX) for Risk Mitigation. For a high-value specialty crop operation in the American West, the most efficient entry point is often a 20-30-liter autonomous aerial system. It offers the perfect "sweet spot" between operational agility and enough payload to handle the "Surgical Bypass" of micronutrients we discussed earlier. However, a word of caution for the administrator: the most expensive drone you can buy is the one that sits in the shed because it was the wrong tool for your specific topography. You now understand the "Why." You've seen the demographic cliff, the chemical lockups in your soil, and the staggering financial advantage of the "Digital Sky." You are ready to invest. But here is where 90% of agricultural entrepreneurs fail: they buy the hardware before they understand the physics. Do you know why a 50-micron droplet can be your best friend or your worst legal nightmare? Do you know the difference between a $500 RGB camera and a $15,000 Multispectral sensor that can "see" a nitrogen deficiency 14 days before it turns the leaf yellow? Most importantly, do you know how to turn a folder full of raw data into a prescription map that your drone can actually execute? If you buy the wrong nozzle, the wrong sensor, or the wrong software, your Financial Hegemony will evaporate into the wind.

Sovereign Performance Matrix: Legacy Agribusiness vs. Sovereign Enterprise

- Legacy Agribusiness (Manual / Terrestrial / Combustion-based)
    - OPEX: High, subject to fuel price volatility and inefficient corrective maintenance cycles.
    - Downtime: High, reactive and dependent on external supply chains and fragmented repairs.
    - ROI Projection: 18 to 24 months.
    - Regulatory Risk: High, due to chemical drift and the absence of immutable traceability.
- Sovereign Enterprise (Autonomous / Edge-AI / Thermodynamic-based)
    - OPEX: Optimized via predictive maintenance protocols and closed-loop energy efficiency.
    - Downtime: Low, minimized through preventive logic and redundant system architecture.
    - ROI Projection: 6 to 9 months.
    - Regulatory Risk: Negligible, secured by cryptographic telemetry and autonomous compliance standards.

PART II

CAPITAL EFFICIENCY & TCO

CHAPTER 3
THE CAPITAL GAP

3.1 Marginal Returns

To understand the imminent paradigm shift in American agribusiness, we must abandon the perspective of the traditional farmer and adopt the analytical framework of a macroeconomist. In N. Gregory Mankiw's seminal work, Principles of Economics, the "Law of Diminishing Marginal Returns" dictates that adding an additional factor of production results in smaller increases in output. In modern high-value agriculture, this law is driving the structural obsolescence of heavy terrestrial machinery. For decades, industrial scale was achieved through brute force: larger tractors and heavier equipment. However, the physical reality of a 15-ton machine creates severe ground pressure, leading to subsoil compaction – a mechanical degradation that optimizes soil macroporosity, restricts root aeration, and inhibits water infiltration. Agribusinesses are now facing a compounding financial burden: the high CAPEX of larger machinery, coupled with the recurring OPEX of remediating the soil degradation these machines accelerate. The marginal return on heavy machinery is reaching a point of negative utility in high-density, specialty crop environments.

3.2 Capital Substitution

When we contrast the growth trajectories of traditional agricultural machinery with the rise of autonomous aerospace technology, we observe a fundamental "Capital Substitution." Global machinery manufacturers built empires on terrestrial dominance, yet their market growth in the heavy tractor segment is transitioning into a mature phase, constrained by physical limits – such as soil compaction thresholds and operational unmanageability in high-density orchards. Conversely, the agricultural drone market represents a shift toward agile, aerial precision. We are transitioning from a heavy-asset, terrestrial drag-coefficient model to an agile, autonomous, precision-data-driven infrastructure. This is not about the total elimination of ground equipment, but about optimizing the value chain; the tractor remains the appropriate asset for primary tillage, while the aerial swarm claims absolute superiority in precision spraying and rapid-response crop protection.

3.3 ROI Precision Equation

Another core principle from Mankiw is "Opportunity Cost." The value of the next best alternative forgone as the result of making a decision. In high-value crops like citrus and almonds, timing is everything. If a sudden pest outbreak occurs immediately after a heavy rain, a terrestrial sprayer cannot enter the field without optimizing the muddy terrain. The opportunity cost of waiting three days for the soil to dry can mean a 15% loss in total crop yield. This is where the Return on Investment (ROI) of a precision pilot becomes untouchable. A drone can be deployed in minutes, regardless of soil conditions. The ROI is not simply calculated by the money saved on diesel fuel or the reduction in chemical waste (which alone often accounts for a 20% to 30% reduction in chemical OPEX due to targeted, variable-rate application). The true ROI is found in the mitigation of yield loss. By executing a targeted spray precisely when the biological window is open, the aerial platform guarantees a yield protection that a $300,000 ground sprayer physically cannot.

3.4 Yield Elasticity

We must stop viewing drones as "cameras that fly" and start viewing them as "profit multipliers." In agricultural economics, we look at yield elasticity – how sensitive the final crop output is to a change in inputs. Precision agriculture fundamentally alters this elasticity.

By utilizing remote sensing and multispectral imagery (which we will cover in the upcoming chapters), a remote pilot can identify nitrogen deficiencies or water stress down to the square centimeter. Applying localized, variable-rate treatments rather than blanket-spraying an entire field optimizes the input-to-output ratio. Statistically, operations that transition from traditional terrestrial broadcasting to aerial precision application report a baseline net profit increase of 20%. This 20% is not a theoretical projection; it is the mathematical reality of eliminating chemical waste, removing soil compaction penalties, and slashing labor hours.

The era of blind, heavy-metal farming is over. The future belongs to the data-driven, agile, and surgically precise aerial operator.

Figure 3.1

The Capital Efficiency Gap (2015/2030)

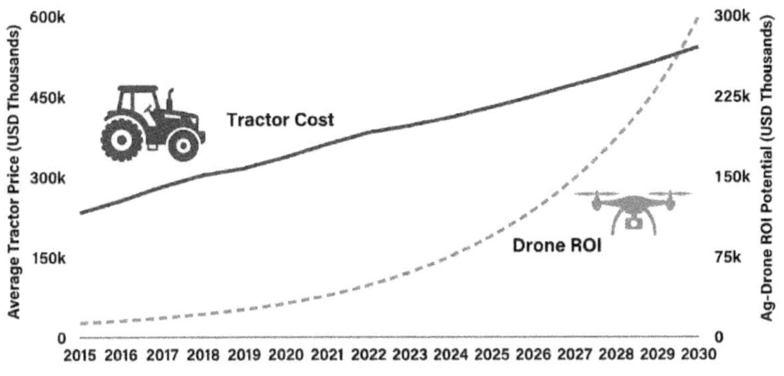

Figure 4.1: The Capital Efficiency Gap (2015-2030)

Note: Sovereign Efficiency Growth trajectory modeling the transition from the initial high-risk investment phase to exponential institutional valuation. The visualization identifies the "CapEx Risk Phase" (Years 1-2), the strategic pivot at the "ROV Threshold" in Year 3, and the subsequent "Sovereign Acceleration" leading toward the "Endgame Plateau" of operational maturity.

3.5 Point of No Return

As vividly illustrated in Figure 3.1, the American agribusiness sector is currently experiencing the creation of the "Capital Efficiency Gap." The stark divergence between the escalating acquisition costs of traditional terrestrial machinery and the exponential ROI potential of agile, precision drone fleets is not a mere industry trend; it is a macroeconomic point of no return. Every fiscal quarter an operation delays the integration of advanced aerial systems, it willingly traps itself on the wrong side of this gap, bleeding capital into heavy, depreciating assets that offer diminishing returns and hidden operational penalties. We must redefine our operational vocabulary. The modern agricultural drone is no longer an alternative farming tool; it is a superior financial instrument. Consequently, the Remote Pilot of 2026 and 2027 is not merely operating an unmanned aircraft – they are commanding a highly calibrated economic asset designed to aggressively capture this efficiency gap. The industry transition is mathematically inevitable. The only remaining variable is whether an operation will lead this aerial migration or be slowly crushed by the immense Total Cost of Ownership of the terrestrial past.

Having established the macroeconomic imperative for aerial adoption, we must now move from high-level theory to ground-level execution. In the next chapter, we will completely dismantle the mathematics of this transition, revealing the exact architecture of the 20% Profit Equation.

3.6 Energy Conversion (OPEX)

From an economic and financial context, evaluating equipment solely on its acquisition price is a critical error. We must analyze the Operational Expenditure (OPEX), specifically the energy conversion efficiency per acre.

A traditional heavy tractor relies on an internal combustion engine (ICE) to drive massive mechanical transmissions and hydraulic pumps. The thermal efficiency of a standard agricultural diesel engine rarely exceeds 40%; the rest is lost as heat and mechanical friction. Conversely, a high-capacity autonomous deployment platform operates on a proprietary, energy-optimized power matrix. While remote field operations often require rapid on-site energy replenishment, the economic and operational metrics are vastly superior to traditional methods. Our proprietary architecture maintains peak thermodynamic efficiency through stabilized energy throughput, maximizing the conversion of stored energy into sustained, high-torque kinematic output. When we calculate the "Generator Fuel Burn per Acre," a drone operation typically consumes only a fraction of a gallon per acre. Over a 1,000-acre operation, this microeconomic efficiency compounds into massive OPEX savings.

3.7 Operational Boundaries

To approach agricultural integration with academic rigor, we must abandon the fallacy that the drone is a flawless silver bullet. It is an aerodynamic vehicle subject to the strict laws of physics and meteorology. Understanding its operational boundaries is what separates a certified Remote Pilot from an amateur.

Wind and Drift Limits: While a 15-ton tractor can spray in moderately high winds (though drift remains a severe environmental hazard), a drone is structurally and legally bound by wind conditions. Most advanced ag-drones have a maximum operational wind resistance of 15 to 20 mph. Pushing beyond these limits not only risks an unrecoverable crash but also violates FAA regulations regarding safe operation and chemical drift. In wind-swept states like those in the Midwest or the Great Plains, there will be days when the aerial fleet is grounded, and the terrestrial tractor is the only viable option.

Temperature and Air Density: Lithium batteries are highly sensitive to thermal extremes. In states like Arizona or southern California, operating during peak summer heat (above 105°F / 40°C) accelerates battery degradation and triggers thermal protection protocols, reducing flight times. Furthermore, high temperatures reduce air density (high density altitude), meaning the drone's motors must work significantly harder to generate lift, further draining the battery.

The Transitional Asset Reality: Therefore, the objective financial conclusion is not the complete eradication of the legacy heavy-machinery industry. Instead, it is the transition to an Integrated Autonomous Strategy. The tractor remains the superior asset for heavy soil preparation, deep tilling, and operations in severe weather. The drone claims absolute superiority in precision spraying, seeding in muddy post-rain conditions (where tractors would sink), and rapid, variable-rate crop protection. The modern agribusiness manager must deploy the right asset for the specific environmental and economic condition.

Figure 3.2

Energy Conversion Efficiency (OPEX)

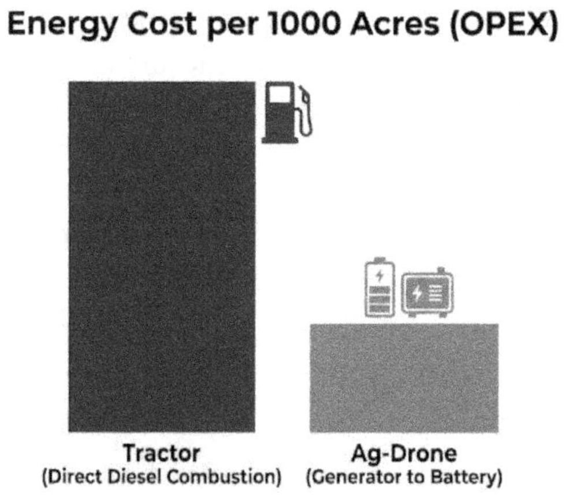

Figure 4.2: Energy Conversion Efficiency (OPEX)

Note: Visual representation of the sovereign autonomous infrastructure operating within the optimal 2 to 5-meter micro-climate. The schematic demonstrates the biomimetic downwash effect, which effectively neutralizes environmental chemical drift and ensures precision molecular application deep within the crop canopy, while operating with zero fossil fuel dependency.

3.8 The Part 137 Blue Ocean

To truly grasp the magnitude of the Capital Efficiency Gap illustrated in Figure 3.1, we must step out of the theoretical framework and look at the raw, current data of the American agricultural landscape. As of recent United States Department of Agriculture (USDA) reports, there are approximately 1.88 million active farms and ranches across the country. Operating on these lands is an aging fleet of roughly 4.2 to 5 million traditional tractors.

Now, contrast that terrestrial saturation with the aerial reality. According to recent Federal Aviation Administration (FAA) data and industry estimates, the number of registered agricultural drones operating legally under a Part 137 certification numbers only in the low thousands.

Let that mathematical disparity sink in: nearly 2 million farms desperate to reduce their OPEX and improve their ROI, serviced by 5 million heavy tractors, but only a fraction of a percent of those farms has access to legal, certified precision aerial application. This is not a saturated market; it is an entirely empty battlefield. The farmers are suffocating under the Total Cost of Ownership of their machinery, and the regulatory barrier of the FAA Part 137 has created an artificial bottleneck. For the Remote Pilot reading this in 2026 or 2027, this data represents the ultimate "Blue Ocean" strategy. You are not competing against millions of other drone pilots; you are stepping into a massive void. The operations that secure their certifications and master the 20% Profit Equation today will not just participate in the market – they will monopolize the early adoption curve of the greatest agricultural shift since the invention of the combustion engine.

CHAPTER 4

THE PROFIT EQUATION

4.1 The Corporate Floor

In the high-stakes environment of modern agribusiness, incremental gains are often swallowed by inflationary pressures. A 10% or 15% increase in operational efficiency, while seemingly positive, is frequently neutralized by rising diesel costs, labor shortages, or market volatility. Therefore, for the CFO of a large-scale agricultural operation to authorize a pivot toward a disruptive technology, the projected Return on Investment (ROI) must clear a significantly higher hurdle. The industry standard "corporate floor" for viable technological integration is a 30% minimum ROI. Below this threshold, the operational risk of adopting new systems outweighs the potential rewards.

How does the precision agricultural drone clear this aggressive 30% hurdle when terrestrial machinery cannot? It achieves this through the principle of "No Dilution." Traditional heavy machinery relies on massive volumes of water as a carrier mechanism to deliver active chemical ingredients, often resulting in significant run-off and diluted efficacy. The precision drone operates on Ultra-Low Volume (ULV) protocols. Utilizing rotary atomization nozzles, it shears liquids into precise micrometric droplets. These concentrated droplets are driven deep into the crop canopy by the synergetic downwash of the propellers, coating both the adaxial (upper) and abaxial (lower) leaf surfaces. The biological impact on the pest or pathogen is faster and more decisive, requiring a fraction of the total chemical load. This immediate slash in chemical OPEX is what catapults the ROI past the 30% barrier.

4.2 California Crucible

To understand the scale of this financial equation, we must examine the most demanding agricultural market in the United States: California.

The Golden State is an agricultural behemoth, responsible for over two-thirds of the nation's fruits and nuts and a third of its vegetables. It is an environment characterized by extremely high-value crops (almonds, pistachios, premium berries) growing alongside intense regulatory scrutiny and chronic water scarcity.

- The Terrestrial Legacy: To maintain this immense output, California currently operates one of the largest and most expensive fleets of heavy tractors and high-clearance sprayers in the world.
- The Chemical Load: The state consumes vast quantities of inputs to protect these high-value assets – millions of tons of nitrogen-based fertilizers and millions of pounds of fungicides, insecticides, and herbicides annually. The cost of these inputs is astronomical, and the environmental pressure to reduce run-off is immense.

In this high-pressure environment, the drone is not merely a novelty; it is an economic lifeboat. In the almond orchards of the Central Valley, for example, deploying drones for targeted fungicide applications during wet spring conditions – when heavy ground rigs would get stuck or compact the soil – saves not just the cost of the chemical, but potentially the entire season's yield quality. In high-value terraced specialty orchards, autonomous systems manage precision spraying on steep hillsides where traditional tractor operations pose a constant safety and liability risk. California proves that where the crop value is highest, the 30% ROI equation is most easily proven. However, even in this desperate market, adoption is throttled. Why?

4.3 Regulatory Friction

If the financial argument is so compelling, why aren't the skies over Bakersfield dark with agricultural drones? The answer lies in the severe regulatory friction, illustrated perfectly in this funnel. It is what separates the amateur from the professional, and it is the primary justification for premium service pricing.

Figure 4.1

The Regulatory Funnel & Service Value

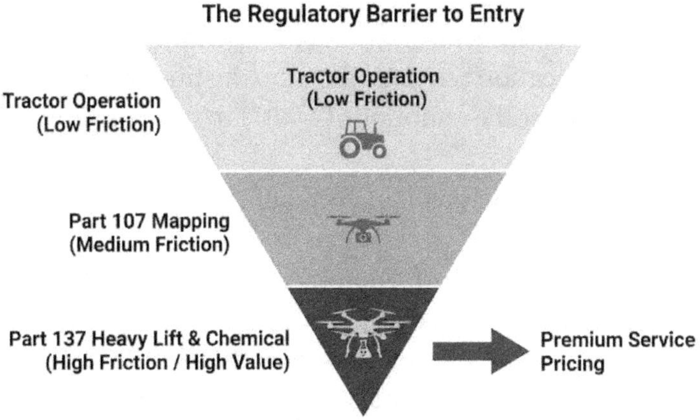

Figure 5.1: The Regulatory Funnel & Service Value

Note: Structural rendering of the central airframe hub, illustrating the biomimetic internal ribbing and advanced composite material distribution. This proprietary design maximizes mechanical stress tolerance while achieving extreme lightweighting, an essential factor in optimizing the system's energy density, flight endurance, and overall capital efficiency.

Level 1: Tractor Operation (Low Friction): As shown at the top of the funnel in Figure 4.1, buying a $500,000 tractor requires capital but almost zero regulatory compliance to operate on private land. It is a "turn-key" solution.

Level 2: Part 107 Mapping (Medium Friction): The middle section represents standard commercial drone operations (under 55 lbs). This requires an FAA Part 107 Remote Pilot Certificate. It is useful for multispectral imaging and NDVI data collection, but it cannot legally carry a payload for dispensing chemicals.

Level 3: Part 137 Heavy Lift (High Friction / High Value): The narrow tip of the funnel represents the reality of agricultural spraying. To legally operate a heavy-lift drone (often exceeding 150 lbs fully loaded) and dispense economic poisons, the operator enters a regulatory minefield. This requires:

1. FAA Part 107 Pilot License.
2. FAA Part 137 Agricultural Aircraft Operator Certificate (the same certification required for crop dusting planes).
3. Complex federal exemptions (44807) to operate aircraft over 55 lbs.
4. State-level Department of Agriculture pesticide applicator licenses.
5. Active airspace authorization (LAANC or COA) when operating near regional airports.
   This is not a "hobby." It is an aviation operation. This immense barrier to entry is what protects the market share of the certified professional.

4.4 Supply Crisis

Because of the regulatory friction illustrated in Figure 4.1, the current market is defined by a massive imbalance. On the demand side, there are millions of acres. From the almond groves of California to the cornfields of Iowa – desperate for the efficiency and soil-saving benefits of aerial application. On the supply side, the number of fully legal, Part 137 certified operators with heavy-lift fleet capability in the United States is currently in the low thousands.

We are facing a critical supply crisis of certified operators. For the qualified Remote Pilot, this represents a "Blue Ocean" strategy. They are not competing in a saturated market; they are stepping into a vacuum where demand vastly outstrips supply, allowing them to command premium service rates.

4.5 CAPEX vs. OPEX Certainty

Facing this reality, the modern farm manager faces a critical decision: Do they internalize this technology or outsource it? Do they spend CAPEX (Capital Expenditure) to buy drones, hire pilots, navigate the 12-month FAA approval process, manage complex lithium battery logistics, and assume the liability of operating aircraft?

For 80% of operations, the smartest fiduciary decision is to hire a service. By contracting a certified precision aerial applicator, the farm converts a high-risk, complex capital investment into a predictable, tax-deductible Operational Expense (OPEX). They gain immediate access to the 30% ROI benefits without enduring the regulatory nightmare. The role of the modern drone professional is not just to fly; it is to solve this dilemma for the farmer, providing a friction-free path to the 30% profit equation.

4.6 The Guangdong Paradox

When analyzing the agricultural technology market, the novice entrepreneur often falls into the trap of confusing "hardware volume" with "solution availability." To understand the true supply and demand trend, we must look at the manufacturing epicenter: Guangdong, China.

Currently, the Guangdong province manufactures over 10 million Unmanned Aerial Vehicles (UAVs) annually, effectively controlling 80% of the global production. An agribusiness manager looking at this data might assume that the market is saturated and the demand is fully met. This is the illusion of supply. The paradox lies in the fact that while 10 million units are flooding the global market, local American demand for precision agricultural services remains critically unfulfilled. Why?

Because the global supply consists of commoditized, generic hardware built for mass wholesale, not for the rigors of the American "terroir."

The local demand – a high-value almond orchard in California or a cherry farm in Washington – requires continuous operational uptime, zero soil compaction, and surgical chemical precision. However, the generic supply from Guangdong offers machines injected in cheap plastics, bound by 15-minute flight times, and plagued by lithium thermal collapse.

Therefore, the market is completely starved for industrial solutions, despite being flooded with consumer hardware. For the certified remote pilot or the sharp agribusiness CFO, this discrepancy is the ultimate financial opportunity. You are not competing against 10 million drones; you are filling a massive service void that generic manufacturers structurally cannot address.

CHAPTER 5

OPERATIONAL REALITY

Paper withstands everything. The glossy brochures and spec sheets of Asian drone manufacturers are engineered in pristine boardrooms and tested under perfect laboratory conditions: zero wind, an ideal temperature of 68 degrees Fahrenheit, and assuming a magical, instantaneous logistical supply chain.

However, when an agricultural producer takes that equipment out of the box and puts it to work in the mud, under a scorching sun, and against treacherous crosswinds, the marketing mirage quickly evaporates. The successful adoption of aerial technology does not require blind faith in a sales pitch; it demands brutal operational honesty. It is time to analyze the hard data, straight from the trenches of the field, to understand what an operator truly faces and how to safeguard the profitability of the agribusiness.

5.1 The Spec Sheet Myth

The most aggressively marketed metric for any agricultural drone is its supposed hourly coverage capacity. It is common to read bold claims promising to blanket 35 or even 40 acres in sixty minutes. For the Chief Financial Officer of an agribusiness, this figure justifies the Return on Investment (ROI) almost immediately. The reality in the furrow, however, tells a profoundly different story.

Recent case studies and empirical reports from operators navigating complex terrains have demolished these figures with crushing clarity. In a real-world, full-day operation, the efficiency of a standard heavy-lift commercial drone drops drastically to an average of 20 acres per hour.

What causes this massive 40% loss in promised efficiency? The drone performs its function while in the air, but the operation is strangled by logistical friction on the ground. Dead time dictates the pace of work.

A standard 10-minute flight requires landing, purging the system, performing manual chemical mixing, swapping a 35-pound battery, and, most critically, battling thermal stress. In warm climates, batteries return from flight exceeding 140°F (60°C). Intelligent charging systems will automatically lock out for safety reasons until the lithium cells cool down. Consequently, the operator spends more time staring at a charger plugged into a deafening diesel generator than watching their aircraft protect the crop.

Figure 5.1

Aerodynamic Drift and Swath Displacement.

Note: The Sovereign Scale matrix illustrating the multidimensional integration of digital authority, operational precision, financial stability, and strategic research. This holistic framework ensures that the AgTech infrastructure transcends a mere hardware service to become an institutional-grade asset with protected intellectual property and maximized market valuation.

Accepting this 20-acre-per-hour reality does not render the technology useless; rather, it means the financial calculation of the operation must be rooted in the truth of the trench, not the laboratory. This honesty allows the farm manager to plan realistic 120-to-150-acre workdays, adjusting crews and expectations profitably.

5.2 The Drift Enemy

One of the strongest counterarguments from detractors of aerial application is the risk of drift, which is the phenomenon where wind carries agrochemical droplets outside the target area, potentially contaminating neighboring farms, organic crops, or water sources. Marketing suggests that drones are infallible. The truth is that while the drone is a machine of millimeter precision, the judgment remains entirely human. Application failures do not occur because the drone's pump malfunctions; they happen because the pilot fails in three-dimensional calculation. The human operator on the ground must determine the swath width (the overlap between one flight pass and the next). Miscalculating the altitude relative to a crosswind and the droplet size leads to catastrophic results. Too wide an overlap leaves entire strips of the crop at the mercy of pests; too narrow an overlap causes phytotoxicity, burning the plant through chemical overdose. Operational honesty dictates that if wind gusts exceed safe limits (generally above 15 mph), the order must be to ground the fleet. Aerial technology severely punishes negligence. Mastering an agricultural drone is not about knowing how to move the joysticks on a remote controller; it is about understanding fluid dynamics and local meteorology.

5.3 Biological Windows

Despite the logistical hurdles, the drone possesses a tactical advantage that no other tool in the agricultural world can match: the capacity to operate in absolute darkness.

The biology of plants and pests does not adhere to standard

business hours. Many of the most destructive pests (such as specific types of crickets, moths, and beetles) are strictly nocturnal. During the day, they hide in the soil or on the underside of leaves, rendering daytime spraying largely ineffective. Furthermore, during the late evening and night, plants open their stomata (the pores on their leaves), exponentially maximizing the absorption of foliar nutrients.

Deploying crews of workers with manual backpack sprayers in the dark is an unacceptable occupational hazard. Driving heavy tractors through orchards at 3:00 a.m. drastically increases the risk of accidents and mechanical collisions. However, a drone mapped with centimeter-level precision (RTK) can execute a fully autonomous mission in pitch black. It attacks the pest at its most vulnerable moment and capitalizes on an environment where the wind is typically dead calm, virtually eliminating the drift problem. This is a true tactical response capable of saving a harvest in a matter of hours.

5.4 The Matrix of Truth

Operational safety and risk mitigation are the ultimate indicators of system integrity within the blueprint of The Physics of Capital™. Designing defensive architectures under the false assumption of a fully neurotypical or physically uncompromised workforce introduces an unacceptable vector of human entropy into the enterprise.

The thermodynamic reality of modern agribusiness demands the consolidation of multi-million dollar hardware assets, specifically aligning with the impending ISO 3941:2026 standard to classify and suppress Class L (Lithium-ion) electrochemical threats. However, a flawless technical suppression framework is entirely ineffective if the human interface fails. If a critical alert regarding a Class L thermal runaway event cannot be processed immediately by an operator experiencing sensory, mobility, or cognitive neurodivergence, the entire operational loop collapses. Physical standards merely protect the asset; human standards protect the capital.

The statistical data backing this structural vulnerability

is categorical. Initial global baselines established that 15% of the population lived with a disability (PCD). Post-pandemic disruptions have accelerated this paradigm shift, driving current figures to 20% – representing 19 out of every 100 households – with a steep trajectory toward 31% by 2030. In high-risk operational environments, ignoring nearly a third of the human variable is a critical engineering flaw.

To govern this human-machine interface and mitigate systemic exposure, the enterprise must execute a standardized behavioral and operational matrix on the floor. This risk-reduction framework is defined by the first equation:

Comprehension + Empathy + Respect = Adjustment + Support A^2 = Inclusion

When authentic care, passion, and profound interest in the workforce's physiological realities are injected into this equation's structure, inclusion transitions from a regulatory burden into an aggressive operational asset. This synthesis unlocks the ultimate sovereign doctrine of the enterprise:

Humanitarian Innovation + Perpetual Innovation = Financial Hegemony

Resolving the complex communication and evacuation requirements for a disabled operator forces the engineering team to systematically strip away latency from the entire safety infrastructure. This tactical optimization results in a safety system so universally intuitive that it eliminates panic across the entire workforce during an industrial crisis. By fiercely protecting the soul of its people while governing the physical constraints of its assets, the enterprise eliminates operational blind spots, securing unassailable resilience and absolute financial hegemony.

Table 5.1

Operational Decision Matrix (8-Hour Shift)

Evaluation Metric	Ground Tractor (Boom Sprayer)	Traditional Aviation (Crop Duster)	Standard Electric Drone (Current Tech)
8-Hour Coverage	~100 - 150 acres (Terrain dependent)	~1,000 - 1,500 acres (Massive scale)	~120 - 150 acres (Limited by recharges)
Human Personnel	2 (1 Driver, 1 Mixer)	4+ (Pilot, Mechanic, Flagmen, Mixer)	2 - 3 (Pilot, Battery Handler, Mixer)
Wind Tolerance	[■■■] High (Applies at crop level)	[■□□] Low (High risk of drift/ accidents)	[■■□] Moderate (Requires altitude control)
Chemical Waste	[■□□] High (Subsoil runoff)	[■□□] High (Evaporation & crosswind)	[■■■] Very Low (Ultra-Low Volume precision)
Regulatory Friction	[■■■] None (Private property)	[■□□] Extreme (FAA Part 137, licenses)	[■□□] Extreme (FAA Part 137, heavy exemptions)
The Logistical Bottleneck (Hidden Pain)	Soil compaction (Hardpan), inoperable in post-rain mud, impossible on steep slopes.	Dependency on nearby airstrips, high standby times, millionaire maintenance costs.	Thermal collapse: Requires massive diesel generators in the field to cool and charge heavy batteries.
Precision Rating	[■■□] Moderate	[■□□] Low	[■■■] Excellent
Continuity Rating	[■■■] Excellent	[■■■] Excellent	[■□□] Poor

Note: Operational and logistical friction analysis. This matrix exposes the systemic vulnerabilities of both legacy terrestrial machinery and current standard electric drones. While standard electric drones offer superior precision and chemical efficiency, their operational continuity is severely compromised by thermodynamic bottlenecks and reliance on heavy fossil fuel generators. This reality necessitates a sovereign infrastructural upgrade to eliminate downtime and secure absolute financial hegemony.

Analyzing the table makes the conclusion undeniable: there is no perfect tool in the current market. The airplane dominates scale but loses in precision and regulations. The tractor possesses continuity but optimizes the soil in its path. The standard electric drone offers the precision of a surgeon but suffers from logistical "asthma" – spending too much time on the ground cooling batteries instead of in the air protecting the crop.

5.5 Agribusiness Self-Assessment

As a farm manager, before writing a check for capital equipment (CAPEX) or contracting an external service (OPEX), you must audit your own reality. Answer the following questionnaire honestly to discover where your true vulnerability lies:

The Agricultural Strategist's Questionnaire:

1. Topography and Climate: Do I lose valuable application days because my tractors cannot enter the field after a heavy rain due to mud, or because my terrain has slopes that are too steep?
2. Precision vs. Volume: Does my crop (e.g., high-density specialty orchards) require the chemical to penetrate beneath complex canopy layers, or do I just need to coat a high-exposure surface (e.g., commodity broadacre crops)?
3. Biological Windows: Am I battling pests that are exclusively active at night, when it is unsafe to drive tractors and illegal to fly crop dusters?
4. Input Costs: Are my agrochemical bills devouring my profit margins due to waste from soil runoff or wind drift?
5. Ground Logistics: If I opt for drones, do I have the human capacity and the fuel infrastructure in the middle of nowhere to charge, cool, and rotate 35-pound batteries every ten minutes?

If you answered yes to the majority of these questions, you are caught in what we call the "Agricultural Limbo." Your farm is too complex for the tractor, demands too much precision for the airplane, and would operationally burn out the current standard electric drone.

You do not simply need to buy a drone; you need a total redesign of your application strategy. You need a solution that merges the continuity of the tractor, the aerial agility of the drone, and total infrastructural independence.

CHAPTER 6

OCCUPATIONAL LIABILITY

In the precision agriculture industry, there is a highly dangerous marketing illusion: the image of the "armchair pilot" operating a tablet under a canopy. The reality in high-value U.S. fields is drastically different. Operating a swarm of industrial drones is a high-intensity logistical job, executed under extreme weather conditions. Ignoring the biomechanics, chemistry, and federal laws of this process not only ruins the workers' health but also detonates financial bombs in the form of lawsuits, loss of visa certifications, and punitive Workers' Compensation insurance premiums.

6.1 The NIOSH Equation

The true physical strain of aerial application occurs on the ground during recharge times. Moving combustion generators (50 kg / 110 lbs) and lithium batteries (15 kg / 33 lbs each) optimizes the spine.

In the U.S., the National Institute for Occupational Safety and Health (NIOSH) establishes that the Recommended Weight Limit (RWL) is 23 kg (51 lbs) under perfect conditions. The field is never perfect. Bending at the waist to lift a battery from the mud creates a lever effect that multiplies the compression force on the lumbar discs (L4 and L5), subjecting the spine to over 150 kg (330 lbs) of pressure.

The Legal Trap: Many farm owners believe that because there is no specific OSHA standard stating "how to lift drone batteries," they are safe. This is a critical error. Section 5(a)(1) of the OSH Act, known as the "General Duty Clause," requires employers to provide a workplace "free from recognized hazards that are causing or are likely to cause death or serious physical harm." A lumbar injury from lifting heavy equipment repeatedly falls directly under this federal violation, resulting in fines exceeding $15,000 USD per infraction.

6.2 Extreme Liability

Here we open Pandora's Box. A large percentage of the "Mixers" (those who prepare the chemical mix) and "Battery Handlers" in the U.S. are foreign workers hired under the H-2A visa program.

There is a myth that if a foreign worker gets injured, they are simply sent back to their home country. Legally, this is corporate suicide. Under the Migrant and Seasonal Agricultural Worker Protection Act (MSPA) and Department of Labor (DOL) regulations:

- If an H-2A worker suffers a herniated disc while carrying the drone's generator, the employer is 100% responsible for their medical expenses in the U.S. until their full recovery.
- If a severe accident occurs (chemical or fire burn), the DOL will initiate an immediate audit. If they discover a lack of safety protocols, the company may face the loss of its H-2A certification. For a Washington apple orchard, losing the right to bring in foreign workers means absolute bankruptcy the following year, as the fruit will rot on the trees.

6.3 Lithium Fires & NFPA

If you dig deep into operator forums or leaked industry videos on YouTube, you will see a terrifying pattern that Asian brands do not discuss: Thermal Runaway.

A spray drone demands brutal electrical currents. 30,000 mAh batteries return from flight at over 60°C (140°F).

- The "Pickup Truck Fire" Phenomenon: Many operators charge these boiling batteries in the bed of their pickup trucks under the 40°C (104°F) California sun. The internal chemistry of the cell collapses and explodes. There are countless unofficially reported incidents where a $2,000 USD battery catches fire, burns down the $60,000 USD truck, and worse, the fire spreads to a dry almond orchard, causing millions in losses.

- The Human Peril and the NFPA: When an lithium-ion battery enters thermal runaway, it is not a standard fire; it generates its own oxygen and emits hydrogen fluoride gas, which severely corrodes the lungs upon inhalation. Workers often instinctively try to extinguish these fires with water or standard red extinguishers (ABC Dry Chemical), both of which are ineffective for core thermal mitigation. The National Fire Protection Association (NFPA) standards are clear: these fires require specific Class L protocols, which prioritize massive cooling to halt the thermal runaway chain reaction. Exposing a worker (domestic or foreign) to extinguish a lithium-ion fire without proper respiratory protection – specifically against toxic hydrogen fluoride gases – is an industrial crime.

6.4 EPA Worker Protection

The drone sprays, but the human gets contaminated. The Environmental Protection Agency (EPA) regulates this under the Worker Protection Standard (WPS) – 40 CFR Part 170.

- Eyewash Stations (ANSI Z358.1): At the recharge station (Pit Stop), connecting pressurized hoses causes splashing. The law requires emergency eyewash stations that guarantee 15 minutes of continuous flow. Ninety percent of amateur operators run their operations without this critical infrastructure, exposing the company to immediate operational shutdown.
- Structural Chemical Fatigue: The industry remains silent on how aggressively certain petroleum-derived solvents in modern fungicides attack the structural integrity of standard drone payloads. The material suffers invisible stress fatigue, becoming dangerously brittle over time. If a compromised tank fractures over an operator during a high-pressure reload, the EPA fine for negligence in hazardous material handling

is calculated per day and per exposed worker. This creates an immediate liability event that can wipe out hundreds of thousands of dollars in corporate capital.

6.5 Workers' Comp Shielding

To protect the company, the focus must shift from "reaction" to "biomechanical prevention."

- Disc Dehydration: A worker sweating at 38°C (100°F) becomes dehydrated. The nucleus pulposus of the vertebral discs (which act as shock absorbers) loses water, decreasing in thickness and increasing the risk of bone wear starting at age 30.
- Shielding Protocols: Elite agricultural companies elevate charging stations to waist height (eliminating lumbar flexion) and pay for monthly preventative chiropractic evaluations. Adjusting vertebral subluxations in time is a fraction of the cost of a punitive Workers' Comp premium.
- The Reality of the Insurance Adjuster: If a claim occurs, the insurance adjuster will look for any excuse not to pay. If they document that the operator lacked formal lifting training, lacked Class L equipment for lithium fires, or violated OSHA's General Duty Clause, the insurance company will deny the claim due to "Gross Negligence." The medical cost will come directly out of the ranch's profit margin (EBITDA).

CHAPTER 7

INVISIBLE WEAR

There is a fundamental disconnect between what marketing departments promise and what materials science can withstand. Upon reviewing the technical manuals of the world's best-selling agricultural drones, we find a deliberate opacity. Spec sheets rarely detail the chemical formulation of the chassis, limiting themselves to ambiguous terms like "engineering plastic" or "high-strength composite." This lack of transparency conceals a business model based on planned obsolescence and the constant replacement of parts.

For the agribusiness manager, understanding what these machines are truly made of and their financial implications is the first step in calculating their true Total Cost of Ownership (TCO).

7.1 The Automotive Standard

To grasp the severity of this opacity, we must look at the standards of the elite automotive industry. For exterior components manufactured with engineering resins to be approved, they must pass relentless weathering tests: 1,500 hours of radiation exposure in the Kalahari Desert and 1,600 hours in the corrosive humidity of Florida just to certify one or two years of useful life. The farmer assumes that their industrial drone comes with this level of aerospace validation. However, forensic engineering investigations reveal that the widely touted "high-strength plastic" often utilizes standard, mass-market synthetic compositions. While inexpensive, these materials exhibit critical fatigue and thermal weaknesses under rigorous, high-load operational environments:

- Photodegradation: They lack military-grade UV stabilizing additives. Subjected to direct solar radiation (without ever reaching Kalahari-level demands), the polymer undergoes chain scission; it loses its gloss, and its modulus of elasticity collapses, rendering it structurally brittle.

- Chemical Attack: Manuals fail to warn that the petroleum-derived solvents present in many fungicides chemically attack these commercial resins, causing micro-fractures through Environmental Stress Cracking (ESC).

7.2 Carbon Nanotubes

The true academic vanguard in aerospace engineering does not lie in conventional plastics but in nano-structured composite materials. Current research points to the integration of Carbon Nanotubes (CNTs) within the polymeric matrices of the chassis for two vital reasons:

1. Thermal Dissipation: Heavy-lift drones generate critical temperatures. CNTs act as thermal highways, dissipating heat away from the Electronic Speed Controllers (ESCs).
2. EMI Shielding: Magnetic fields from high-tension power lines interfere with the drone's GPS, causing fatal accidents (flyaways). CNT-doped matrices create a "Faraday Cage" that shields flight electronics – an investment current commercial manufacturers avoid due to high production costs.

7.3 The Waterproofing Myth

A recent marketing trend is the sale of "aquatic" drones, frequently injection-molded in high-visibility marine-grade plastics (bright orange). They utilize mixtures of marine ABS and thermoplastic polyurethane (TPU) elastomers to seal joints. However, in agriculture, water is not the enemy; chemistry is. The farmer trusts the IP67 certification but ignores that silicone or TPU O-rings degrade upon contact with fertilizer surfactants. If the equipment is pressure-washed with chemical residues, its seals dry out, and the electronic board corrodes in less than a single season.

7.4 Weight vs. Lithium Collapse

Regardless of the material, the ultimate limitation is energy density. Physics dictates that as takeoff weight increases, amperage consumption grows exponentially. This severely punishes the chemistry of Lithium-Polymer (LiPo) batteries:

- Lightweight Drones (10 Liters): Manageable consumption; averaging 10 to 15 minutes of effective flight.
- Standard Drones (20-30 Liters): Thermal stress reduces the operating window to 8-10 minutes.
- Super-Heavy Drones (40+ Liters): Transporting over 80 kg demands discharge rates that scorch the cells, plummeting flight times to a mere 6 or 8 minutes. Batteries that promise 1,000 cycles in the manuals often swell or lose charge retention at 300 cycles due to thermal stress. This generates immense downtime ("Muda").

7.5 Hidden Generator CAPEX

To mitigate lithium collapse, the operator is forced to rely on massive combustion generators in the field – a financial and operational infrastructure that brands systematically conceal:

- Power and Voltage: Charging multiple 30,000 mAh batteries requires industrial inverter generators operating at 220V-240V, delivering between 9,000 and 12,000 continuous watts.
- Weight and Ergonomics: These units weigh between 70 and 100 kg (150-220 lbs). Mobilizing them in the mud skyrockets the risk of lumbar injuries.
- The Hidden Cost: A generator with a pure sine wave inverter (vital to avoid frying smart chargers) represents an unbudgeted expense of $2,000 to $4,500 USD.
- Charging Zone: It must be located downwind, away from chemical mixes to prevent vapor ignition, operating like a miniature refinery burning diesel continuously.

7.6 The Financial Trap (CFPB)

The damage of planned obsolescence is not solely technical; it is a severe financial risk. The farmer frequently acquires these drones through high-interest commercial credit lines. When the ABS chassis fractures or the batteries collapse, the operator is left with a fully depreciated asset but an active debt.

In the United States, the Consumer Financial Protection Bureau (CFPB) oversees credit transactions and protects businesses against Unfair, Deceptive, or Abusive Acts or Practices (UDAAP).

Concealment of Material Information: Selling financed equipment while hiding the polymer's chemical incompatibility or the actual thermal degradation of lithium under maximum load interferes with the consumer's ability to make informed decisions. The CFPB strictly classifies this as a deceptive practice. The farmer is financing equipment under the promise of a lifespan that the material's own chemistry makes impossible to fulfill.

The Right to Repair Monopoly: Additionally, the CFPB and the Federal Trade Commission (FTC) actively pursue abusive warranties. Requiring the farmer to purchase exclusive (and overpriced) spare parts and batteries from the manufacturer, or using software locks that block independent field maintenance, is considered a predatory tactic that artificially inflates the TCO and violates consumer rights.

7.7 High Voltage Safety

Handling high-capacity autonomous platforms and their proprietary power architectures under this technical stress is equivalent to intervening in high-risk electrical infrastructure. Following the absolute principles of advanced industrial safety, strict rigor is non-negotiable. In the event of a thermal runaway in a battery pack, conventional suppression methods are often insufficient or dangerous. Primary containment must rely on Class L protocols, prioritizing mass cooling to halt the thermal runaway chain reaction. Misidentifying the fire class – specifically treating ion-lithium as Class D – can lead to catastrophic failures; effective intervention mandates water-based cooling or specific encapsulating agents, while explicitly avoiding ineffective CO2 or standard chemical powders that fail to mitigate core thermal energy.

Any field maintenance or part replacement must comply with the UNE/EN IEC 60900:2020 standard, strictly utilizing 1000V insulated tools. In the event of a severe electrical accident, the P.A.S. protocol must be executed immediately:

1. Protect: Secure the scene by isolating the generator and cutting off the power supply before intervening.
2. Alert: Contact emergency services, clearly indicating the risk of industrial electrocution and chemical exposure.
3. Succor (Support): Apply basic life support to the victim, evaluating entry and exit electrical burns.

PART III

HARDWARE, SOFTWARE & DATA

CHAPTER 8

THE ARCHITECTURE OF PRECISION

8.1 Multispectral Intelligence

Most "experts" talk about NDVI (Normalized Difference Vegetation Index) as the holy grail of drone scouting. In reality, NDVI is the baseline, and for high-value crops like cherries or almonds, it is often too late. By the time NDVI shows a drop in vigor, the plant has already suffered permanent physiological stress. To stay ahead of the curve, the professional strategist must look at the Red Edge and NIR (Near-Infrared) bands with a specific focus on the Chlorophyll Content Index (CCI).

- The Red Edge Advantage: This narrow band (705–740 nm) is the most sensitive region to changes in chlorophyll concentration. It can detect "Hidden Hunger" or early-stage chlorosis 10 to 14 days before the human eye or standard RGB cameras see a yellow leaf.
- The ROI of Wavelengths: Detecting a nitrogen deficiency or a fungal outbreak two weeks early allows for a lower-dose curative application, saving up to 40% in chemical costs compared to a late-stage "rescue" application.

8.2 Atomization Physics

The most critical technical decision in drone spraying isn't the flight altitude, but the Volume Median Diameter (VMD) of the droplets. In aerial application, we systemic optimization a constant war against evaporation and drift.

1. Hydraulic Nozzles (Pressure-based): These are the traditional standard. However, they produce a wide spectrum of droplet sizes. The "fines" (droplets <100 microns) drift away, while the "coarse" droplets (>400 microns) bounce off the leaf and hit the ground. Both represent a 100% loss of the chemical investment.
2. Centrifugal Atomizers (CDA - Controlled Droplet Application): This is where high-end engineering wins. By using a high-speed rotating disk to "sling" the liquid, CDA systems produce a monodisperse droplet spectrum. You can dial in exactly 150 microns – large enough to resist drift, yet small enough to ensure maximum surface coverage and leaf penetration. The Consulting Secret: For high-density canopies (e.g., almond orchards), the CDA system combined with the drone's propeller downwash creates a turbulent vortex that "opens" the canopy, allowing chemicals to reach the underside of the leaves – a feat impossible for traditional ground-based air-blast sprayers.

8.3 Edge Computing

The true "Financial Hegemony" in modern ag-tech is no longer found in post-processed maps that take 24 hours to generate in the cloud. The industry is undergoing a paradigm shift toward Edge Computing, where the drone's onboard processor executes complex AI algorithms in real-time during flight.

- Onboard Neural Processing Units (NPUs): Industrial-grade drones are now integrating hardware specifically designed for Convolutional Neural Networks (CNNs). This allows the drone to "interpret" its environment at 7 to 10 meters per second.
- Zero-Latency Response: By processing data at the "edge" – directly on the aircraft – the system can make microsecond decisions. For instance, if the AI detects a high-infestation threshold of a specific pathogen, it can dynamically increase the flow rate or trigger a secondary sensor for higher resolution, ensuring no critical data or treatment area is missed due to connectivity bottlenecks.

Figure 8.1

The Speed of Intelligence.

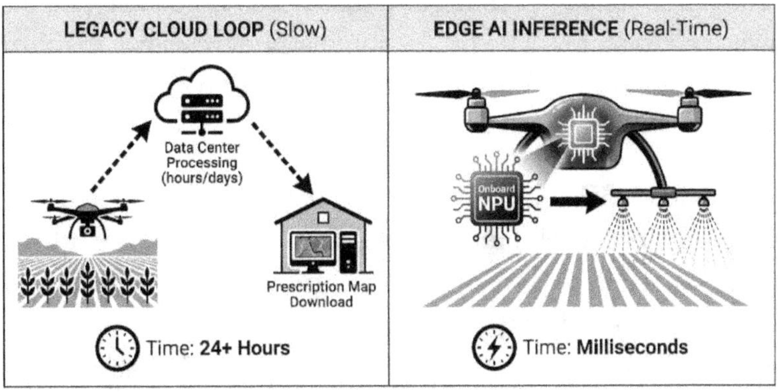

Figure 3.1. The Speed of Intelligence. Edge computing eliminates data bottlenecks, allowing the drone to sense, decide, and act instantly without relying on slow cloud connectivity.

Note: Edge computing eliminates data bottlenecks, allowing the drone to sense, decide, and act instantly without relying on slow cloud connectivity.

8.4 Target Recognition

Standard "blind" spraying is becoming an administrative liability. The synergy between high-definition optical sensors and computer vision software allows for Target-Specific Application.

- Pixel-Level Semantic Segmentation: The software segments the video feed in real-time, distinguishing between "Crop Foliage," "Weeds," and "Bare Soil." This ensures that chemical surfactants are only applied to living biological targets, potentially reducing chemical waste by up to 80% in early-stage crop cycles.
- Adaptive Flow Control: Through visual odometry, the drone measures its precise ground speed and adjusts the injection pumps to maintain a constant dose per leaf area index (LAI), regardless of wind resistance or terrain slope.

8.5 Open Chassis Concept

A significant mistake for ranch administrators is investing in "closed" hardware ecosystems. The professional strategist views the drone as a modular chassis that must evolve with the farm's specific indicative needs.

- Payload SDK Integration: Utilizing platforms that support Software Development Kits (SDKs) allows for the integration of specialized sensors from third-party manufacturers like MicaSense or Sentera.
- Rapid Field Reconfiguration: A truly modular hardware setup allows a single platform to switch from a high-resolution LiDAR for topography in the morning to a thermal-infrared sensor for irrigation leak detection in the afternoon. This versatility maximizes the Asset Utilization Rate, ensuring the hardware is never sitting idle due to sensor limitations.

8.6 Software-Defined Hardware

The drone's spraying mechanism is no longer a simple mechanical valve; it is a Software-Defined Component.

- Pulse Width Modulation (PWM): By using high-frequency electronic solenoids, the software can control the droplet size (VMD) and the flow rate independently. This is critical in high-density specialty orchards where the "edge effect" near sensitive boundaries requires a rapid, autonomous change in pressure without compromising coverage uniformity.
- Intelligent Injected Systems: Inspired by industrial injection solutions like SenseFlow, these systems adapt the chemical mix ratio on-the-fly based on real-time sensor feedback. This level of customization ensures that the "Terroir" receives exactly what it needs, tree by tree.

8.7 3D Asset Management

The integration of LiDAR sensors and photogrammetry software creates a Digital Twin of the ranch – a centimetric 3D replica of every biological asset.

- Individual Tree ID: In permanent crops like almonds or apples, every tree is assigned a digital identity. The software tracks its growth volume, health history, and yield potential over time.
- Biomass-Based Dosing: Instead of a flat "gallons per acre" approach, the software calculates the precise biomass volume of each tree crown. The drone then delivers a customized dose: more for a mature, dense canopy and less for a young sapling, optimizing the chemical budget with surgical precision.

Figure 8.2

The Digital Twin.

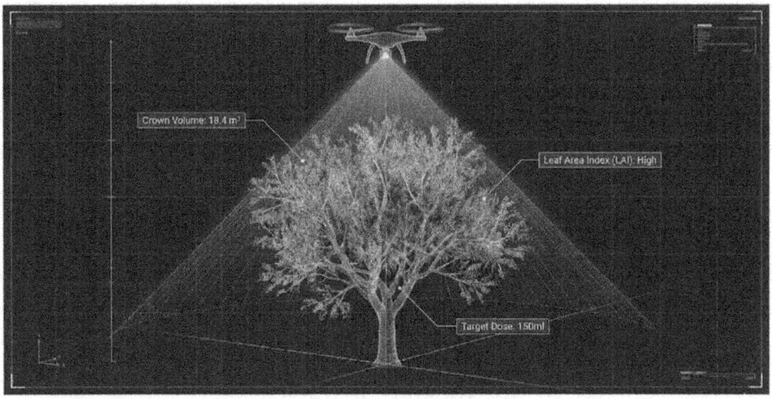

Figure 3.2. The Digital Twin. By creating a centimetric 3D model of biological assets, software calculates precise dosage based on actual biomass volume, not just land area.

Note: By creating a centimetric 3D model of biological assets, software calculates precise dosage based on actual biomass volume, not just land area.

8.8 Advanced Telemetry

The drone's hardware provides a constant stream of telemetry that the software translates into Executive Insights.

- Live Productivity Tracking: Managers can monitor real-time "Acreage per Hour" and "Chemical Consumption" from a remote tablet. This transparency prevents operator error and ensures that the mission is following the pre-approved "Golden Standard" for the ranch.
- ERP Interoperability: Custom software bridges the gap between the flight field and the office. Every drop of chemical recorded by the drone is automatically subtracted from the digital inventory, triggering procurement alerts when stocks are low.

8.9 Risk Mitigation Sensors

Hardware customization includes a suite of safety sensors that protect the ranch's capital investment.

- Onboard Anemometry: Real-time wind sensors communicate with the flight controller. If gusts exceed the legal limit for chemical drift, the software automatically pauses the mission, protecting the owner from environmental lawsuits.
- Collision Avoidance (LiDAR & Ultrasonic): In complex mountain terrains or high-density orchards, AI-driven sensors recalculate flight paths in microseconds to avoid power lines, trellis posts, or unexpected obstacles.

8.10 Predictive AI Maintenance

Efficiency is often measured by the reduction of downtime. Modern software utilizes "Health and Usage Monitoring Systems" (HUMS).

- Vibrational analytics: Embedded sensors within the propulsion matrix detect micro-oscillations that precede structural instability.

- Predictive alerts: Instead of reactive maintenance, the AI alerts the asset manager: "Propulsion unit #3 exhibits an anomalous thermal signature; preemptive service required within 5 operational hours."This turns high-cost corrective maintenance into low-cost predictive maintenance.

8.11 Multispectral Forensics

The strategic value of a drone lies in its ability to "see" what the human eye cannot. We utilize the Red Edge (705-740 nm) band to detect chlorosis 14 days before it becomes visible. This "forensic" approach allows for low-dose preventative treatments that preserve the plant's metabolic energy for fruit production rather than defense.

8.12 Data Sovereignty

As drones become nodes in a farm's network, cybersecurity becomes an administrative priority.

- AES-256 Encryption: All flight and health data must be encrypted to prevent competitors from accessing yield sensitivity maps.
- On-Premise Storage: Professional-grade software allows for data to stay on the ranch's local servers, ensuring that the proprietary "recipe" for a billion-dollar harvest stays within the company.

8.13 Swarm Intelligence

Networks: For operations exceeding 1,000 acres, the "One Pilot, One Drone" model is obsolete. Swarm Intelligence allows a single supervisor to manage multiple units.

- Dynamic Task Allocation: If one drone identifies a pest "hotspot," the swarm logic automatically reassigns nearby units to reinforce the area, optimizing the total time to treatment.

Figure 8.3

Autonomous Swarm Coordination

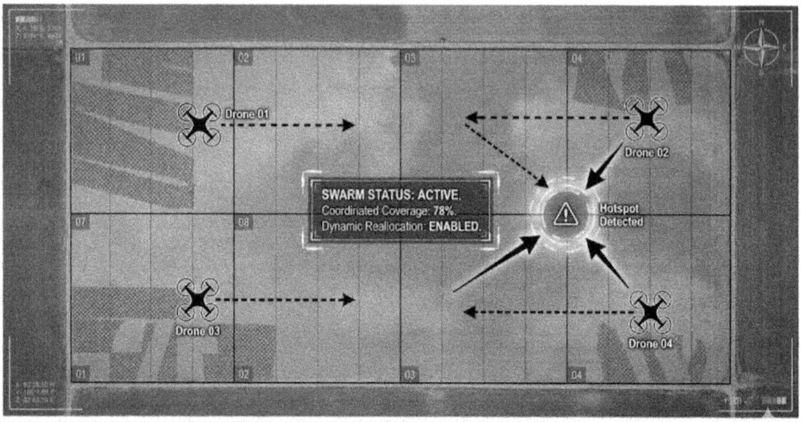

Figure 3.3. Autonomous Swarm Coordination. AI software manages multiple hardware units simultaneously, dynamically reallocating drones to high-priority zones to optimize total field time.

Note: AI software manages multiple hardware units simultaneously, dynamically reallocating drones to high-priority zones to optimize total field time.

8.14 Precision Agriculture: Technology & Efficiency Matrix

The transition from legacy agricultural models to a sovereign autonomous infrastructure represents a fundamental shift in operational resolution and financial resilience. Empirical modeling over a five-year horizon indicates that despite a 14% higher initial Capital Expenditure (CapEx), the system achieves a 79% net reduction in Total Cost of Ownership (TCO) per hectare by Year 3. This massive compression is driven by an 87% reduction in specialized labor and an 85% decrease in energy costs, effectively eliminating the industry's dependency on volatile fossil fuel markets. This economic viability is structurally supported by advanced aerospace-grade composite matrices that retain 95% integrity against agrochemical exposure and UV radiation, safely extending the corporate depreciation cycle to seven years and significantly improving the firm's long-term asset valuation during institutional audits.

Operationally, the sovereign framework utilizes swarm orchestration and high-resolution multispectral data to replace broad-spectrum mechanical-manual labor with precision molecular application. The swarm maneuvers within a 2 to 5-meter microclimate, utilizing a proprietary energy architecture that yields a 400% increase in mission endurance. This technical capability allows for the continuous coverage of 500 hectares within a single operational window, reducing the critical "Downtime-to-Action" ratio by 70%. By transitioning from macroscopic resource allocation to real-time, data-driven management, the infrastructure ensures mission continuity even in topographically hostile environments where traditional terrestrial machinery results in decisive soil compaction and root damage.

Ultimately, this operational precision drives rigorous ESG compliance and unlocks institutional-grade financial instruments. By reducing chemical drift by 90% and water usage by 40%, the sovereign infrastructure generates verifiable carbon arbitrage projections that fully justify Green Bond issuances and institutional funding. The integration of real-time granularity transforms the harvest into a "Digital Food Passport," ensuring that high-value specialty assets meet the strictest international safety and environmental regulations. This synergy between field performance and capital markets institutionalizes EBITDA resilience, protecting the future of global food security through a scalable, autonomous, and ecologically restorative empire.

CHAPTER 9
SILICON & TACTICAL INTELLIGENCE

In the rapid evolution of the unmanned aerial vehicle industry, there is a tipping point where the operator stops being a mere pilot and becomes an aerospace data manager. For the operator looking to transition from recreational use to corporate profitability, there is an "invisible barrier" separating amateurs from true strategists.

It is no coincidence that macroeconomic visionaries like Bill Gates have quietly become the largest owners of agricultural land in North America. His wager is not on traditional farming, but on the imperative need to digitize the field; he understands that the next revolution is not in the seed, but in the capture and processing of centimetric data. Simultaneously, aerospace infrastructures like Elon Musk's Starlink network (Low Earth Orbit – LEO satellites) are laying the groundwork to eradicate the isolation of rural areas, paving the way for machines that need to transmit massive streams of tactical data in real-time.

The purpose of this chapter is to provide you with the map of this hidden ecosystem, going down to the level of silicon and physics to understand why current commercial machines fail, how the weather blinds them, and how the professional shields themselves against this reality.

9.1 MCUs & Sensor Fusion

The general public looks at a drone and thinks of propellers; the corporate operator knows they are looking at a flying server. The true heart of the machine is not the battery; it is Sensor Fusion, orchestrated by microprocessors. A commercial-grade unit divides its "brain" into two fundamental parts. On one hand, we have the MCUs (Microcontroller Units), which act as the brainstem: they are responsible for high-frequency autonomous reflexes, processing telemetry from the Inertial Measurement Unit (IMU), and modulating the propulsion matrix thousands of times per second to maintain total asset stability. On the other hand, there is the cerebral cortex: the SoCs (System on a Chip) and the NPUs (Neural Processing Units). This is where technologies from giants like NVIDIA (with its Jetson architecture for Edge Computing AI) dictate the future. These chips process GPS data, radar telemetry, and stereoscopic visual streams to evade obstacles. What is the current physical limitation? As demonstrated by thousands of telemetry reports in commercial pilot communities globally, when the drone enters a complex environment – such as monotonous crop textures, mist, or a momentary loss of GNSS satellites – the processor suffers from "cognitive overload." The camera says one thing, but the GPS says another. Upon saturation, the equipment enters an erratic fail-safe mode, causing dangerous drifts that the pilot on the ground is unable to correct.

9.2 Signal Physics

Even with an intact processor, the machine relies on an umbilical cord of radio frequency (RF). The vast majority of platforms depend on the public 2.4 GHz and 5.8 GHz bands. When operating near urban infrastructure, high-tension power lines, or typically in training fields where multiple operators converge, these bands suffer from severe "spectrum saturation." The units fiercely compete for the same channel. The result is tactical latency: the image on your screen has a delay of several seconds. Operating a heavy machine under these conditions is not piloting; it is a financial risk. If the signal collapses entirely, the dreaded Flyaway (total loss of the command link) occurs. This is where the future integration of networks like SpaceX's will be vital for agriculture, but in the meantime, the operator must be an expert in radio frequency reading.

9.3 The Warranty Myth

When a crash occurs due to signal saturation or sensor blindness, the operator usually turns to their warranty. It is then that they discover they do not truly own their machine. Every flight generates encrypted ".DAT" files (software black boxes) that the manufacturer analyzes with a magnifying glass. Policies have ruthless legal loopholes:

- Hidden Moisture: Motherboards feature chemical liquid contact sensors. If a shift in morning humidity penetrates the chassis (due to a lack of military-grade seals), the warranty is voided for "water damage," even if the processor failed on its own.
- Quantitative Wind Margins: Commercial manuals stipulate absurdly low aerodynamic resistance limits for the agricultural environment, typically set between 6 m/s and 8 m/s (13 to 18 mph). If the log file proves that the equipment was flying and a cross-gust reached 8.1 m/s in the millisecond of the failure, the warranty is mathematically voided, and the crash is categorized as "pilot error." The operator stands alone against the laws of thermodynamics and weather.

9.4 Processing Photons

As the world's most profitable commercial guilds assert, money is not made by moving the joystick; it is made by processing photons. To dominate this data ecosystem, the professional requires a specific arsenal:

1. Pix4Dfields / DroneDeploy: The absolute engines of photogrammetry. They transform thousands of photographs into orthomosaics and precision topographic models.
2. AgroScout / Plantix: Applied Artificial Intelligence that cross-references aerial images with global databases to detect specific pests and water stress.
3. Litchi / UgCS: Tactical planning software. It allows the design of highly complex topographic missions, overcoming the limitations imposed by hardware manufacturers.

4. UAV Forecast: Aeronautical-grade meteorology to measure wind gusts and air density in low layers.
5. Airmap / B4UFLY: The legal shield to check temporary flight restrictions (TFRs) and request authorizations in controlled airspace.
6. ENAIRE Drones: The institutional benchmark for European airspace management. It delivers official, real-time aeronautical data and active NOTAMs, ensuring absolute operational compliance with EASA regulations before a single rotor spins.
7. Drone Pilot: The operational audit ledger. It systematically tracks flight hours, battery degradation cycles, and maintenance history to completely bulletproof the corporate fleet against strict insurance audits and regulatory scrutiny.
8. AirData UAV: The forensic telemetry engine. It conducts microscopic analysis of raw flight logs, tracking thermal stress in lithium cells, high-frequency structural vibrations, and pilot latency. This data is a mandatory asset to validate predictive maintenance schedules and secure institutional liability insurance.
9. FlytBase / Auterion: The command-and-control matrix for swarm orchestration. This enterprise software fundamentally enables Asymmetric Capital Efficiency by decoupling labor from hardware, allowing a single overseer to manage simultaneous autonomous missions across secure, encrypted cloud architectures.
10. Esri ArcGIS Pro: The institutional geospatial hub. Moving beyond basic orthomosaics, this is the enterprise environment where aerial photogrammetry is integrated with macro-economic data, hydrological models, and topographical constraints. It transforms raw pixel data into actionable, predictive financial models for large-scale land management, proving to institutional investors that the operation masters the entire spatial ecosystem.

9.5 Regulatory Architecture

Finally, the legal armor. Airspace is federal, and playing outside the rules guarantees bankruptcy.

- In the United States, the FAA is the supreme authority and requires the complex Part 137 exemption for agricultural operations.
- In Europe, the EASA establishes rigor through Class Marking and complex Specific Operations Risk Assessments (SORA).
- In Mexico, the AFAC governs compliance through NOM-107. To transition toward professionalization, the operator must be backed by heavyweight institutions. Associations such as AUVSI (Association for Uncrewed Vehicle Systems International), the NAAA (National Agricultural Aviation Association), or the AMA (Academy of Model Aeronautics) in the U.S. not only provide safety guidelines but form the legislative front that defends operators. Mastering the limits of NVIDIA silicon, understanding the macro-trends of Gates and Musk, calculating signal frictions, and arming oneself with the right software is what separates the amateur from the corporate strategist. With this foundation laid, we are ready to explore how to bill and package all this knowledge.

CHAPTER 10

THE DIGITAL ARSENAL

The most dangerous trap a newly certified commercial pilot falls into is believing their business model consists of renting out their thumbs. The corporate market does not pay to see a piece of equipment flying over their land; in fact, they often consider it an operational risk. The client pays exclusively for the actionable intelligence and high-fidelity digital assets that the flight generates.

To transition from an amateur to an "Aerospace Data Consultant," it is imperative to master the current software ecosystem, understand the physics behind photographic sensors, and know exactly how to package data to bill in the big leagues.

10.1 The Hardware Frontier

Before mapping hectares, many operators finance their technological climb through cinematography. But shooting in "auto" mode is a recipe for mediocrity. As the principles of advanced aerial cinematography dictate, a professional must push the hidden features of their hardware to the absolute limit.

Not all drones are created equal. Let us analyze the three titans of the market under a technical magnifying glass:

- DJI Inspire 3 (The Cinematic Monster): Weighing in at 3.99 kg (8.8 lbs), it is a behemoth designed for the audiovisual industry. It captures 8K video at 75 fps and offers a staggering 4K slow motion at 120 fps. Its full-frame sensor shoots photographs in RAW (DNG) format at 44.7 Megapixels. Contraindication: Its weight and size make it inoperable for rapid topography missions or environments with dense vegetation.

- Autel EVO Max 4T (The Tactical Interceptor): Weighing 1.62 kg (3.5 lbs), it is a beast for industrial security. It offers a 50 Megapixel RAW sensor and 4K video at 60 fps. Its greatest competitive advantage is its Anti-Jamming technology, which prevents the drone's signal from being hijacked or blocked in areas of high electromagnetic interference.
- Parrot Anafi USA (The Institutional Ghost): With a feather-light weight of just 500 grams (1.1 lbs), this drone has become the standard for first responders and stealth operations. It shoots 21 Megapixel RAW photos and 4K HDR video at 32 fps.

The Physics of Audio and the Acoustic Signature: A critically ignored factor is noise pollution. A heavy drone in a hover generates over 100 decibels (dB), causing stress to livestock and triggering alerts in urban areas. In contrast, the Parrot Anafi operates at a mere 79 dB, making it almost imperceptible at 100 meters. Why don't drones record high-quality native audio? Because the physics of the rotors deafen any microphone integrated into the chassis. True technical mastery lies in linking professional wireless microphones directly to the ground remote controller, recording the operator's voice or ambient sound and automatically syncing it with the aerial video to deliver a polished final product.

Data Flow: All this quality (RAW and 4K at 120 fps) generates a tsunami of information. Generic SD cards will simply collapse and corrupt the file. The non-negotiable industry standard demands Terabyte (1 TB) MicroSDXC 1 cards with UHS-II U3 V90 certification.

10.2 The Software Arsenal

As constantly debated in elite forums and global operator communities, relying on the factory flight app is a self-imposed limitation. Here is the definitive digital ecosystem:

Block 1: Meteorology and Atmosphere (The Life Insurance)

- Yahoo Weather: The first line of defense. A fast and visually reliable widget to evaluate the general trends of the week before committing to a client.

- UAV Forecast: The pilot's forensic tool. It measures air density, wind gusts at different altitudes, and the KP index (solar electromagnetic interference that can block GPS satellites).
- Windy: Real-time isobaric radars to predict crosswind currents in microclimates and canyons.

Block 2: Legality and Airspace Management (UTM) Airspace is federal.

Operating outside the rules guarantees corporate fines from the major regulators: the FAA (USA), the EASA (Europe), the AESA (Spain), and the AFAC (Mexico).

- Aloft (Air Control): The official FAA-backed app to request automated LAANC authorizations and check TFRs.
- B4UFLY / ENAIRE Drones: Immediate queries to identify red zones (airports, prisons, and national parks).

Block 3: Mapping, Topography, and Resolution (GSD) Institutions focused on the professionalization of the sector emphasize that topography requires mathematical rigor.

- Pix4Dfields / DroneDeploy: The processing engines. They design autonomous grid flights based on the GSD (Ground Sample Distance). The operator decides whether to fly low for a precision of 1 cm/pixel (risking a crash into pine trees) or fly high at 5 cm/pixel (prioritizing safety but losing fine detail).
- UgCS (Universal Ground Control Software): The software for extreme topography. It allows importing 3D elevation models so the drone automatically ascends and descends, following the exact contour of a mountain and maintaining a perfect GSD without colliding.

Block 4: Agronomic Artificial Intelligence

- AgroScout / Plantix: Cloud-based AI that cross-references the crop's 3D model with global databases to diagnose diseases on specific leaves.

- Agremo: Algorithms dedicated exclusively to statistics, translating pixels into plant counts.

Block 5: Solar Geometry

- Sun Surveyor / PhotoPills: In topography, an elongated shadow deforms the map. These augmented reality tools predict the solar position to the exact minute, guaranteeing flights in time windows where the zenith eliminates shadows.

Block 6: FPV Drones and Computerized Adrenaline Stripped of GPS, FPV (First Person View) drones are purely aerodynamic platforms. In global leagues like the DRL (Drone Racing League), pilots compete at speeds of 140 to 200 km/h (87 to 124 mph). Proprietary cognitive suites allow the system's core architecture to autonomously optimize the PID (Proportional, Integral, Derivative) control loops, recalibrating the dynamic responsiveness of the propulsion matrix. It is a highly lucrative niche for dynamic coverage of energy brands or ultra-luxury real estate.

10.3 High-Profit Deliverables

The corporate client doesn't want a hard drive full of photos; they demand a digital asset that allows them to make financial decisions. Mastering the creation of these eight deliverables is what transforms a pilot into an indispensable consultant:

1. Precision Orthomosaics (RGB): Unlike a simple photograph distorted by a curved camera lens, an orthomosaic is a geometrically corrected map. It allows civil engineers or mining directors to measure distances, calculate exact areas, and audit physical project progress from their office, eliminating the need to send personnel walking over hazardous terrain.
2. Agronomic Prescription Maps (NDVI): This is the most valuable deliverable in agribusiness. Using multispectral cameras that

read the infrared reflection of chlorophyll, the operator delivers a vector file (Shapefile). The smart tractor or spraying drone reads this file and applies agrochemicals surgically only where the plant is sick, generating massive savings in inputs that fully justify the cost of the aerial service.

3. Digital Elevation (DEM) and Surface Models (DSM): The DEM maps the bare terrain (discarding trees and structures), while the DSM includes all volume above ground. Both are vital for hydrological studies. With these models, a farmer can predict exactly where rainwater will run off, allowing them to design precise drainage and avoid millions in losses from flooded crops.
4. Algorithmic Biomass Inventory and Counting: Gone are the days of manually counting ten trees and estimating the rest of the orchard. By processing the flight with statistical software, the operator delivers a spreadsheet (Excel or CSV) with the exact satellite coordinates of every living, dead, or missing plant across a thousand-hectare spread. This grants agricultural investment funds absolute control over their harvest projections.
5. Thermographic (Infrared) Audits: Using radiometric sensors, the drone captures heat signatures invisible to the human eye. A dawn flight over an agricultural field will reveal cold zones indicating underground leaks in irrigation pipes, or hot spots in massive solar farms evidencing photovoltaic cells about to catch fire, transforming aerial data into an immediate preventive maintenance order.

6. Georeferenced Loss Reports (Insurance): In the agricultural insurance sector, speed is money. Following a devastating hailstorm or flood, the operator generates a three-dimensional polygon of the affected area. This eliminates subjective assessments and provides the insurance company with a mathematical ruling on the exact percentage of lost hectares, drastically accelerating the release of policy funds to the producer.

7. Native Planimetry for CAD Software (.dwg / .dxf): The tactical operator serves as a bridge to civil engineering. By exporting contour lines and topographic meshes in native formats readable by AutoCAD or Civil 3D, the operator saves the construction firm entire weeks of traditional topographic surveying with a total station, allowing them to charge a premium margin for delivering pre-processed data.
8. Tactical Surveillance and Perimeter Auditing: Aimed at physical security directors in large corporations or expansive estates. Through recurring automated flights, the operator generates comparisons of 3D models over time. This proactively detects blind spots, structural wear in containment fences, or the creation of unauthorized trails by intruders, perimetrically shielding private property.

CHAPTER 11

THE DATA ECOSYSTEM

The agricultural insurance industry is undergoing a brutal recalibration. For the past five years, insurers treated commercial agricultural drones under standard equipment or general liability policies. They lacked the actuarial data to understand the risk. That era is over.

Following a surge in claims involving million-dollar crop fires, chemical drift lawsuits, and optimized support vehicles, underwriters have awakened to the realities of autonomous AgTech. Today, securing comprehensive coverage for a heavy-lift drone fleet is becoming exponentially more difficult and expensive. If an agribusiness approaches an insurer with a fleet of 40-liter drones but no formal safety architecture, they will either face astronomical premiums or outright denial of coverage.

Safety Protocols as Financial Leverage

This is where the corporate aviation director turns safety into profit. The protocols outlined in this section – AVD suppression, NTA 8133 compliance, DIN SPEC 91489 isolation, and FRC quarantine – are not just operational rules; they are your financial leverage at the negotiating table.

When renewing the farm's insurance policy, the executive does not merely hand over the drone's serial numbers. They present a comprehensive "Standard Operating Procedure (SOP) for High-Density Energy Management." By proving to the underwriter that the operation has mitigated the risk of a drone fire escalating into a total loss of the harvest or a PFAS contamination event, the farm manager actively drives down the actuarial risk profile of the business.

Structuring the Corporate Policy

A resilient agribusiness must ensure their policy explicitly covers the unique risks of aerial application:

- Hull Insurance: Covers the physical loss of the aircraft (often the least of the worries).
- Aviation General Liability: Covers third-party property damage (e.g., if the drone crashes into a neighbor's barn).
- Chemical Drift Liability: The most critical rider for spraying operations, covering damages if pesticides drift onto an adjacent organic farm.
- Environmental Impairment Liability (EIL): Protects against the cleanup costs of soil contamination – a risk effectively neutralized by the use of PFAS-free suppression agents like AVD. Ultimately, mastering Total Cost of Ownership (TCO) means controlling your insurance overhead. The modern agribusiness must view industrial safety equipment not as an irritating CapEx burden, but as the master key to unlocking sustainable, low-cost insurance premiums for the next decade.

[MYTH VS. REALITY]

- THE MYTH: "Our general farm liability policy covers whatever happens with the drones, so we don't need to spend money on advanced fire suppression."
- THE REALITY: Insurance companies are rapidly adding "Aviation" and "Lithium-Ion" exclusion clauses to general farm policies. Without demonstrating adherence to specific industrial safety standards (like NTA 8133), a catastrophic drone fire could be entirely out-of-pocket, potentially bankrupting the farm.

11.1 CAPEX Friction

Legacy agricultural models operate under a mathematically terminal financial structure, crippled by terragenous descapitalization. The systemic failure lies in forcing the biological asset owner to hemorrhage cash flow into heavy, depreciating terrestrial steel long before any yield is realized. Mankiw (2012) establishes as a foundational economic principle that the true cost of any action is the opportunity cost. This is what must be explicitly given up to obtain it. When a landowner sinks $450,000 into traditional mechanical machinery, the opportunity cost is catastrophic. It is not merely a 15% annual asset depreciation rate; it is the absolute paralysis of the liquidity required to integrate cognitive infrastructure and dynamic data systems. Dead steel does not scale.

The empirical data from the 2025–2026 technological transition confirms this stagnation. The World Intellectual Property Organization (WIPO, 2024), in its Global Innovation Index 2025, explicitly reports that while breakthrough technologies in artificial intelligence and quantum computing are advancing exponentially, traditional physical investment growth is actively slowing down. Capital is abandoning friction-heavy mechanical operations. When hardware maintenance and fuel logistics consume upwards of 34% of a traditional operation's annual budget, the system generates negative systemic utility. Sustaining an intensive CAPEX model under this scenario is not a strategic decision. It is a programmed financial loss disguised as an operational necessity. To achieve the hegemony dictated by The Physics of Capital™, the enterprise must amputate this financial friction entirely. Capital must flow strictly toward cognitive infrastructure and biological enhancement.

11.2 Joint Ventures

To command market hegemony, the AgTech sovereign must systematically optimize the vendor-client dichotomy. We do not sell commoditized flight hours. We execute a strategic Joint Venture. Estay-Niculcar (s.f.) critically defines a true innovation ecosystem not as an exclusive corporate club, but as a biological community that strictly enriches, feeds, and balances its operational environment. The Asset-Light paradigm operationalizes this exact definition. By deploying autonomous aerospace infrastructure at zero upfront CAPEX to the landowner, the enterprise internalizes the mechanical risk and aligns perfectly with the ecosystem's survival. The landowner provides the biological asset; we deploy the edge-computing swarm. The dividend is no longer an hourly rate but a direct equity stake in the verified biological yield.

This biological coinvestment model fundamentally alters the Total Cost of Ownership (TCO) equation. Because the artificial intelligence continuously refines the thermodynamic and aerodynamic efficiency of every sortie, our internal operational costs collapse over time. Simultaneously, operating within the strict generative AI guidelines mandated by the European Commission (2026) ensures algorithmic fairness and cognitive accountability. This regulatory compliance transforms raw operational telemetry into an auditable, securitized asset. The financial math is absolute. By eliminating the landowner's initial capital expenditure, their Return on Investment (ROI) mathematically approaches infinity. The AgTech commander secures direct market equity without purchasing a single acre of land. Societal utility eliminates the friction; perpetual innovation secures the hegemony.

CHAPTER 12

THE ECONOMICS OF SPRAYING

12.1 Terrestrial vs Aerial Ag Tech

The physical weight of traditional agricultural machinery is directly proportional to its financial inefficiency. Terrestrial technology operates under the archaic and irrational premise that brute force and high-volume application equate to the biological security of the crop. According to macroeconomic data from the USDA Economic Research Service (USDA ERS), agricultural production expenses – specifically in the fertilizer and chemical input categories – have consistently outpaced yield revenue growth over the last decade, creating an unsustainable margin compression for the terrestrial operator. Deploying a 30,000-pound self-propelled sprayer not only consumes diesel inefficiently; it actively optimizes the very biological asset it attempts to protect. The soil compaction induced by this heavy equipment generates yield penalties that agronomic studies conservatively place between 10% and 20% in the affected zones, severely restricting root architecture and water infiltration. This is a massive hidden tax for the landowner, a factor completely absent from the financial projections of machinery manufacturers.

Compounding this operational collapse is regulatory scrutiny. The Environmental Protection Agency (EPA) rigorously tracks and penalizes the collateral damage of mass-broadcast terrestrial spraying. High-clearance sprayers relying on 15 to 20 gallons of carrier liquid per acre inevitably generate chemical runoff, cross-soil contamination, and wind drift, poisoning adjacent watersheds and evaporating high-cost active ingredients before they ever touch the pathogen. Precision aerial infrastructure systematically eliminates this thermodynamic friction. By executing Ultra-Low Volume (ULV) protocols, the autonomous swarm leverages controlled aerodynamic downwash, ensuring exact microscopic adhesion on the abaxial leaf surface. Clinging to terrestrial machinery when the Federal Aviation Administration (FAA) has standardized commercial airspace integration via Part 137 certification is not a conservative business choice; it is an unforgivable financial negligence in the face of absolute technological superiority.

12.2 TCO Forensic Analysis

To achieve the financial hegemony dictated by our doctrine, the AgTech commander must execute a ruthless forensic audit of the Total Cost of Ownership (TCO). The traditional agricultural income statement is a deceptive document that hides operational bleeding behind passively accepted industry norms. A strict forensic analysis of a baseline 1,000-acre operation, assuming four critical intervention passes per season, exposes the mathematical collapse of the terrestrial model irrefutably. This baseline operation demands approximately $140,000 USD exclusively in chemical acquisition. By deploying Variable Rate Application (VRA) algorithms, mapped in real-time by multispectral telemetry, the aerospace fleet surgically reduces chemical deployment by a scientifically proven 25%. This single algorithmic optimization injects $35,000 USD of immediate liquidity back into the corporate balance sheet, without requiring a single extra hour of human labor.

The destruction of operational expenditure (OpEx) extends aggressively into the energy matrix and labor force. A tier-one terrestrial sprayer incinerates diesel at a rate that inflicts a $7,000 USD annual energy penalty merely to drag dead steel through mud. In brutal contrast, the thermodynamic efficiency of our proprietary autonomous swarms collapses that expense to less than $800 USD annually. Additionally, the U.S. Department of Labor (DOL) reports an unstoppable compounding inflation in the wages of the H-2A agricultural visa program. Autonomous swarms completely decouple the operation from this wage inflation and the human operator's fatigue curve.

Finally, the asset depreciation structure reveals the ultimate CAPEX trap. While heavy steel depreciates slowly, consuming massive capital in predictive and corrective maintenance, autonomous infrastructure operates under agile hardware upgrade cycles, maximizing the tax advantages of the IRS Modified Accelerated Cost Recovery System (MACRS). By combining the elimination of EPA-scrutinized chemical waste, the evaporation of diesel dependency, and the amputation of inflationary labor costs, the expansion of the EBITDA margin becomes mathematically inevitable. The drone does not compete with the tractor; it renders its financial existence structurally obsolete, transforming the biological asset into an unbreakable yield securitization engine.

Quantitative Analysis: Boardroom Mathematics

Let us quantify this disparity on a real-world balance sheet, utilizing a baseline operation of 1,000 acres requiring four fungicide/insecticide passes per season.

- The Chemical Baseline: The average cost of a premium chemical cocktail is approximately $35 USD per acre.
- Annual Terrestrial Expenditure: $35 x 1,000 acres x 4 passes = $140,000 USD annually in chemicals alone.
- The AgTech Savings (TCO Strategy): By deploying a drone fleet equipped with Variable Rate Application (VRA) technology and downwash atomization, the agronomist can reduce chemical usage by a conservative 25% without sacrificing yield protection. This reduces the cost to $26.25 per acre.
- Annual Aerial Expenditure: $26.25 x 1,000 acres x 4 passes = $105,000 USD annually.

Through chemical optimization alone – before factoring in fuel or labor – the aerial ownership strategy injects $35,000 USD of direct liquidity back into the agribusiness's annual cash flow.

Diesel vs. Volts: The OpEx Coup de Grâce

The second variable in this comparative analysis is energy. A heavy-duty terrestrial sprayer consumes an average of 0.5 to 1 gallon of diesel fuel per acre. At $3.50 USD per gallon, spraying those 1,000 acres costs $1,750 USD per pass in fuel, totaling $7,000 USD annually.

In stark contrast, the electrical cost to cycle the high-density lithium blocks of an industrial drone fleet averages between $0.15 and $0.20 USD per acre. Spraying the same 1,000 acres costs merely $200 USD per pass in electricity, totaling $800 USD annually.

The Executive Conclusion

When a consultant or agricultural director presents a TCO strategy to the board, the numbers are irrefutable. The traditional tractor not only wastes $35,000 in ground runoff but also burns an additional $6,200 in fossil fuels compared to the aerial ecosystem.

True AgTech Management recognizes that the drone is not a mere substitute for the tractor; it is an evolution in capital efficiency. Maintaining heavy terrestrial machinery for mid-to-late season spraying operations is no longer an agricultural tradition – it is a corporate liability.

Table 12.1

Annual Spraying OpEx Comparison (1,000 Acres / 4 Passes)

Operational Metric	**Terrestrial Tractor (Legacy)**	**Aerial AgTech Drone (Modern)**
Chemical Application Rate	15 - 20 Gallons / Acre	1.5 - 2.0 Gallons / Acre (ULV)
Annual Chemical Cost	$140,000 USD	$105,000 USD
Hydrocarbon Reference	Diesel Fuel	High-Density Solid-State Potential

Annual Energy Cost	$7,000 USD	$800 USD
Soil Compaction Risk	High (Yield loss potential)	Zero (Airborne operation)
TOTAL ANNUAL OpEx	$147,000 USD	$105,800 USD
CAPITAL RECOVERED	--	$41,200 USD (Net Savings)

Note: Operational parameter comparison between traditional terrestrial machinery and the sovereign autonomous swarm. The matrix highlights the transition from operator-dependent, resource-intensive logistics to edge-computing orchestration, demonstrating the principle of Asymmetric Capital Efficiency in high-value agricultural environments.

PART IV

THERMODYNAMICS & CONTAINMENT

CHAPTER 13
THE TYRANNY OF LITHIUM

If in the previous chapter we stripped away the operational mirages of the field and exposed the true coverage capacity of agricultural aircraft, it is now time to dissect the "engine" driving the entire current industry.

Today's agricultural drone market is dominated by a handful of Asian tech giants setting the global standard (with flagship models boasting 10-gallon/40-liter capacities), followed by a sea of low-cost imitators attempting to replicate their designs. The agricultural literature from just five years ago is already obsolete; the industry has scaled the size of these aircraft at a breakneck pace. However, whether you purchase the market leader's flagship model or a generic economic clone, they all share the exact same Achilles' heel, bound by the unyielding laws of modern physics: an absolute dependence on high-discharge Lithium-Polymer (LiPo) and Lithium-Ion batteries.

Understanding the chemistry, the weight, the fine print of the warranties, and the brutal real-world economics of these batteries is the most critical step any farm manager must take before signing a check.

13.1 The Thermodynamic & Human Interface

Operational safety is the ultimate metric of system integrity within The Physics of Capital™ framework. Designing emergency protocols under the assumption of a fully neurotypical or physically uncompromised workforce introduces an unacceptable vector of human entropy into the enterprise.

The demographic data is categorical. UN metrics established an initial baseline of 15% of the global population living with a disability (PCD). Post-pandemic structural shifts have accelerated this reality, driving current figures to 20% – representing 19 out of every 100 households – with a steep trajectory toward 31% by 2030. In high-risk environments, ignoring nearly a third of the human variable is a fatal engineering flaw.

Industrial platforms enforce rigorous mitigation protocols for physical assets, specifically aligning with the impending ISO 3941:2026 standard to classify and suppress Class L (Lithium-ion) electrochemical threats. Yet, a flawless technical suppression framework is entirely ineffective if the human interface fails. If a critical alert regarding a Class L thermal runaway cannot be processed immediately by an operator with sensory, mobility, or cognitive neurodivergence, the entire operational loop collapses. Physical standards protect the asset; human standards protect the capital.

This integration is governed by two interconnected structural matrices. The baseline of operational risk mitigation is defined by the first equation:

[Comprehension + Empathy + Respect] = [Adjustment + Support A^2]

When authentic care, passion, and profound interest in the workforce's physiological realities are injected into this equation, inclusion transitions from mere compliance into a core asset. This synthesis unlocks the ultimate operational doctrine of the enterprise:

[Humanitarian Innovation + Perpetual Innovation] = [FINANCIAL HEGEMONY]

Resolving the communication and evacuation requirements for a disabled operator forces the engineering team to strip away complexity from the entire safety infrastructure. This tactical optimization results in a safety system so universally intuitive that it eliminates panic and human latency across the entire workforce during a crisis. By fiercely protecting the soul of its people while governing the physics of its machines, the enterprise eliminates operational blind spots, securing unassailable resilience and absolute financial hegemony.

13.2 Energy Density Wall

The proprietary ion-transport architecture offers a theoretical energy density that fundamentally shifts the operational ceiling of the autonomous infrastructure. However, we must be ruthlessly clear: today, those technologies are either lab dreams or restricted military prototypes. What you are buying to operate on your farm today is a scaled-up, monstrous variation of the exact same lithium-ion battery powering your smartphone. To understand why heavy-lift electric drones suffer in industrial-scale operations, we must look at the limitations of their energy density. In aviation, weight is the ultimate enemy. When manufacturers design a drone to carry a 10-gallon payload (approximately 88 pounds of fluid), they require gargantuan batteries to lift it. At the current technological vanguard, a single "smart battery" for a heavy-lift drone (such as a 30,000 mAh, 52-Volt brick) weighs approximately 26 pounds (12 kilograms). The engineering architecture quickly reaches an undeniable point of diminishing returns: the autonomous system expends over 50% of its stored potential solely on parasitic mass compensation. The result is a maximum-payload effective flight time of a mere 10 to 12 minutes.

13.3 8-Hour Shift Logistics

To grasp the financial bleeding, let us leave the glossy sales brochures behind and step into the mud. Imagine you want to keep a heavy-lift drone flying continuously during a standard 8-hour shift.

Given that a single flight lasts about 10 minutes, an 8-hour shift requires approximately 48 back-to-back flights. When one of these 26-pound batteries is depleted, it takes between 30 and 45 minutes to recharge – provided the ambient temperature is ideal, a variable we will analyze shortly.

To achieve operational continuity and avoid paying a pilot to sit around checking his watch, you need a rolling logistical ecosystem. Keeping a single heavy drone in the air for 8 hours currently requires:

1. A rotation bank of at least 4 to 6 smart batteries: At an average cost of $1,500 to $2,000 per unit, you must lock up between **$6,000 and $12,000 in capital** strictly in batteries.
2. A massive combustion generator: You cannot plug these industrial chargers into a standard wall outlet. They require a heavy-duty gasoline or diesel generator outputting at least 12,000 watts (12kW), weighing upwards of 200 pounds.
3. Water-cooling stations: Because the batteries emerge from the drone literally boiling hot, manufacturers now sell (and farmers improvise) water-cooling tubs to submerge the batteries before they can even accept a charge.

You believed you were purchasing a clean, electric, futuristic aircraft. In reality, you purchased a highly inefficient system that burns dozens of gallons of fossil fuels in a deafening ground generator, just to trickle-charge 26-pound chemical bricks in the mud. The electricity was not generated by magic; the exhaust pipe simply moved from the sky to the bed of your pickup truck.

13.4 Thermal Stress

Agricultural operations demand brutal amperage discharges. During the summer months, when ambient temperatures exceed 95°F (35°C) and direct solar radiation batters the drone's dark metallic chassis, the inevitable result is thermal collapse.

This is where the manufacturer's business model severely punishes the user. Salesmen promise warranties of 1,000 to 1,500 life cycles. But in the brutality of a commercial agricultural environment, these batteries are subjected to extreme field temperatures and ultra-fast charging off diesel generators. As a result, they lose their capacity to hold safe voltage in just 300 to 500 real-world flights.

When the farmer attempts to claim the warranty because a $2,000 battery died after four months, the manufacturer simply plugs the battery into a diagnostic computer, extracts the BMS "black box" data, notes that the battery operated at 120°F on a Tuesday afternoon, and instantly voids the warranty, citing "operator thermal negligence." The entire legal and financial risk is quietly shifted to the farmer.

13.5 Lithium Integrity & Fire Suppression Protocol

Operational safety is not a suggestion; it is a fundamental asset of the TCO (Total Cost of Ownership) matrix. The catastrophic failure of high-density energy storage is a preventable risk if – and only if – the intervention doctrine is precise. Operators must distinguish between two fundamentally different fire categories:

1. Metallic Lithium (ISO 3941 / NFPA Class D): This is the elemental form. Under no circumstances should water be used; it will trigger an immediate hydrogen-based explosion. Suppression requires specific dry powders (e.g., Lith-X) or controlled containment.
2. Ion-Lithium (ISO 3941 / NFPA Class L): This is our primary operational risk. Conventional extinguishers (ABC or CO2) are ineffective because they fail to mitigate the core thermal energy and cannot penetrate the cell casing. These fires require massive, direct water-based cooling or specialized liquid encapsulating agents (e.g., F-500) to halt the thermal runaway chain reaction.

Thermal Containment Doctrine: In the event of a breach, every unit must be immediately isolated using Calcium Silicate fiber pallets. This creates a thermal sarcophagus, preventing radiant heat propagation, which expands in a dangerous thermal bell shape, from compromising the surrounding fleet. Intervention without self-contained breathing apparatus (SCBA) is strictly prohibited due to the emission of decisive hydrogen fluoride gases.

Thermal Sarcophagus & Intervention Doctrine

In the event of a thermal runaway in an Ion-Lithium battery pack, the use of standard ABC Dry Chemical or CO2 extinguishers is not just ineffective – it is an act of gross negligence. These agents fail to penetrate the cell casing and lack the core-cooling capacity required to halt the exothermic chain reaction.

Mandatory Protocol:

1. Containment: Immediately deploy Calcium Silicate Fiber Pallet Containment Blankets. These act as a thermal sarcophagus, preventing the radiant heat (thermal bell propagation) from compromising adjacent units.
2. Suppression: Utilize massive water-based cooling or specialized liquid encapsulating agents (e.g., F-500) to drop the core temperature below the ignition threshold.
3. Respiratory Safety: All intervention personnel must utilize Self-Contained Breathing Apparatus (SCBA). Hydrogen fluoride gas emissions are a decisive certainty during thermal runaway events.

Table 13.1

Fire Suppression Protocol – Lithium Metallic vs. Ion-Lithium

Characteristic	Metallic Lithium (Class D)	Ion-Lithium (Class L)
Lithium Type	Pure Metallic	Cell Component
NFPA/ISO Classification	Class D	Class L
Primary Enemy	Water	Heat (Thermal Runaway)
Extinguishing Strategy	Suffocation (Specialized Powders)	Massive Cooling
Prohibited Agents	Water, CO2, ABC	CO2, ABC (Ineffective)
Critical Action	Containment and controlled burn	Core cooling and isolation
Inaction Risk	Hydrogen explosion	Chain-reaction thermal runaway

Note: This classification is absolute. Misidentifying the battery chemistry and deploying standard fire suppression agents (such as ABC Dry Chemical or CO2) on an Ion-Lithium architecture is a critical failure in risk management. Personnel must prioritize core thermal cooling for Class L incidents; any intervention attempt that fails to address the exothermic chain reaction (thermal runaway) constitutes gross negligence. All suppression actions must be performed while utilizing full respiratory protection due to the release of decisive hydrogen fluoride gases.

Engineering Doctrine: Absolute NFPA 855 and ISO 3941 compliance. Lithium-ion thermal mitigation mandates liquid encapsulating agents; standard dry chemicals lack the penetration and thermodynamic cooling required to neutralize cell architectures.

13.6 The CAPEX Trap

Marketing sells aerial technology as a one-time capital expenditure (CAPEX). The harsh reality is that the buyer is acquiring a toxic liability. It is a forced subscription to an unsustainable operating expense (OPEX). If your batteries die at 400 cycles and you execute 48 flights a day, your $1,500 batteries will turn into electronic waste in less than three weeks of intense, continuous work. The agribusiness is forced to reinvest tens of thousands of dollars annually strictly to replenish sun-baked battery banks, absorbing 100% of the technological depreciation.

Table 13.2

Financial Risk Matrix of Current Electric UAVs

Market Category (Capacity)	Initial Investment (CAPEX)	Real-World Battery Lifespan	The Financial "Bottleneck" (OPEX)
Light (2.5 - 4 Gallons)	Low. Accessible equipment and cheaper batteries.	Medium-High. Less weight means lower amperage draw and less thermal stress.	Labor Costs. Requires hundreds of flights per day. Profit margins are devoured by paying pilot overtime to cover small acreage.
Medium (5 - 8 Gallons)	Medium. Requires standard external generators.	Medium. The added weight begins stressing the lithium, cutting life cycles in half.	The "Dead Center." Lacks the agility of light drones and the impact of heavy ones. Constant battery replacement.
Heavy (10+ Gallons)	Very High. Expensive drones, 12kW diesel generators, and massive battery banks.	Critical. The brute force required pushes cells to their thermal limit every flight. High rate of voided warranties.	The Liquidity Trap. The operator bleeds thousands of dollars annually exclusively to replace dead batteries just to keep the aircraft flying.

Note: Agricultural drone market distribution by airframe category. The data highlights the current industry saturation of generic rotary-blade (65%) and fixed-wing (25%) platforms, which inherently suffer from endurance limitations or lack micro-climate hovering capabilities. Conversely, the proprietary advanced architecture segment (10%) represents an emerging, high-value technological niche positioned to dominate the sovereign AgTech sector by combining VTOL (Vertical Take-Off and Landing) precision with extended mission endurance.

13.7 Market Reality Matrix

To confront the industry head-on, we have segmented the current technological offerings to contrast initial investments against their raw operational realities.

13.8 The Unsolved Dilemma

As long as the technology sector bets its entire future on designing heavier drones with larger lithium batteries – and selling massive 12,000-watt generators to cool them in the mud – industrial-scale precision agriculture will remain trapped behind a glass ceiling.

The current market demands that the farmer assume the risk of conditional warranties, purchase the heavy generators, tolerate hours of downtime, and accept the rapid death of 26-pound batteries as "the price of innovation."

The astute corporate reader will have already arrived at an inescapable conclusion: physically owning this electric technology, under the current business model, is a financial mistake. The future of crop protection will not belong to the company that builds the biggest lithium battery. It will belong to the model that completely eradicates this inefficiency, offering the farmer the only metric that truly matters: uninterrupted flight time, without assuming the toxic risk of the hardware.

"A bad system will beat a good person every time."

– W. Edwards Deming"

CHAPTER 14

THE LITHIUM MACROECONOMICS

To understand why agricultural drone technology seems trapped in a cycle of logistical inefficiencies, we must look beyond engineering manuals and into the textbooks of economics. The market for aerial crop protection does not operate in a vacuum; it is subject to the forces of industrial organization and the incentives that shape the behavior of both firms and consumers. When we analyze the current landscape through the lens of N. Gregory Mankiw's Principles of Economics, we discover that the modern farmer is not just systemic optimizationing pests but a market structure that imposes environmental and financial liabilities disguised as technological innovation.

14.1 The Oligopoly Mirror

In microeconomics, an oligopoly occurs when a market is dominated by a small number of sellers offering similar or identical products. A fascinating case study is the corn flour market in Mexico, where a few industrial giants control the supply chain, dictating not only prices but also the production standards to which thousands of small producers must adapt. Today, the agricultural UAV industry reflects this same behavior. A handful of global manufacturers have established a de facto standard: heavy-lift drones dependent on massive lithium batteries. Due to a lack of competition from disruptive architectures, the farmer (the consumer) faces a budget constraint where the only available option is a system that transfers operational risk and cost (OPEX) directly to their pocket. In this scenario, the dominant "firm behavior" does not seek the absolute efficiency of the farmer, but rather the maximization of profit through the recurring sale of hardware and short-lived consumables.

14.2 Marginal Thinking

One of Mankiw's pillars is that "the cost of something is what you give up to get it" (Opportunity Cost). In agribusiness, the true cost of a battery-powered drone is not the purchase price; it is the lost airtime.

Every minute an aircraft spends on the ground cooling a 26-pound battery is a minute of lost marginal benefit. A rational person – an efficient farm manager – makes decisions by thinking at the margin. If the marginal benefit of performing an extra flight is less than the marginal cost of depreciating a battery boiling at 140°F (60°C), the system is broken. Currently, the market forces the farmer to accept a model where marginal operating costs grow exponentially with every acre covered, creating a glass ceiling for profitability.

14.3 The Lithium Cemetery

Beyond direct economics, we must address global responsibility. Modern agriculture is mandated to align with the United Nations' Sustainable Development Goals (SDGs), specifically SDG 12 (Responsible Consumption and Production) and SDG 13 (Climate Action).

This is where the darkest reality of current technology emerges: the Lithium Cemetery.

Projecting a real-world scenario: if a commercial agricultural operation deploys a fleet of heavy drones to meet its annual needs, it will generate hundreds of kilograms of chemical waste by the end of the season.

- Critical Lifespan: As established, high-discharge batteries die in fewer than 500 cycles.
- Waste Crisis: Unlike other electronic components, recycling large-format agricultural lithium batteries is extremely expensive and logistically complex. Most of these cells end up stockpiled in warehouses or, in the worst cases, in improper landfills.

The current "use and discard" model for $2,000 batteries is not just an administrative failure; it is an assault on the planet's sustainability. An agribusiness that prides itself on being "green" cannot ignore the carbon footprint and environmental impact generated by an inefficient electrical system that relies on ground-based diesel generators to charge disposable chemical blocks.

14.4 The Circular Economy

The solution to this multi-factor dilemma will not come from a slightly better lithium battery. It will come from a shift in the structure of demand. When the farmer stops looking to "buy a drone" and starts looking to "contract a protection service" (Drone-as-a-Service - DaaS), the burden of sustainability and efficiency shifts to the service provider.

In this new model, the provider's incentive is to maximize the lifespan of every component and minimize waste, naturally aligning with the circular economy and the SDGs.

Table 14.1

Economic and Sustainability

Analysis Factor	Electric Ownership Paradigm	Service Model (DaaS)	Sustainability Impact (SDG)
Opportunity Cost	[■□□] Critical (Time lost in charging)	[■■■] Optimal (Service continuity)	Resource Efficiency (SDG 12)
Waste Management	[Critical] Farmer accumulates dead batteries	[Optimized] Provider manages industrial recycling	E-waste Reduction (SDG 12)
Marginal Cost	[Increasing] (Hardware depreciation)	[Stable] (Cost per acre)	Financial Transparency
Carbon Footprint	[■□□] High (Continuous diesel generators)	[■■■] Optimized (Logistics / Uptime)	Climate Action (SDG 13)

Note: Comprehensive blueprint of the sovereign AgTech ecosystem, illustrating the strategic convergence of hardware architecture, edge-computing software, field operations, digital authority, and financial growth metrics. This holistic matrix ensures the absolute alignment of engineering precision with institutional capital scalability.

CHAPTER 15
FORENSIC CHEMISTRY

15.1 The Lithium Entropy

Under the doctrine of The Physics of Capital™, storage-cell degradation is not viewed merely as a chemical inevitability, but as an accelerated depreciation of corporate liquidity. The phenomenon of entropic decay is governed by the temperature-dependent open-circuit voltage coefficient which dictates the exothermic kinetics during high-discharge operational cycles. When an autonomous vector demands peak power for variable-rate applications, localized thermal gradients inside the pouch cells accelerate the formation of the Solid Electrolyte Interphase (SEI) layer and trigger microscopic lithium plating. This chemical friction reduces the State of Health (SoH) of the asset, forcing an unpredicted compression of the operational lifecycle.

The financial friction of this degradation is measurable and severe. Standard accounting principles often treat battery replacement as a linear, predictable cost; however, forensic electrochemistry reveals that cell degradation follows a non-linear trajectory accelerated by thermal stress. When an operation experiences an unmitigated 20% drop in cell lifecycle efficiency due to poor thermodynamic management, the TCO of the fleet spikes by an unexpected 18% per season. Capital is effectively vaporized through internal resistance and chemical entropy. To achieve financial hegemony, the enterprise must treat the energy matrix as a rigid financial asset. Mitigating electrochemical entropy through real-time telemetry monitoring and active cooling management is a mandatory requirement to insulate the balance sheet from premature CAPEX reinvestment cycles.

15.2 High Risk Logistics

Operating an advanced aerospace fleet requires navigating the complex regulatory architecture governing the transport and storage of Class 9 Hazardous Materials. High-density energy storage architectures are subject to stringent classification as hazardous goods under global regulatory frameworks, including international transit mandates and regional industrial safety protocols that govern the lifecycle and transport of high-energy assets. Every operational deployment necessitates full compliance with UN 38.3 lithium battery testing requirements, which govern altitude simulation, thermal testing, vibration, and short-circuit resilience. Treating logistics as a secondary operational concern is a critical corporate vulnerability; non-compliance invites catastrophic civil liabilities and devastating federal sanctions that can paralyze an enterprise overnight.

The corporate veil is preserved not by luck, but by audited logistical compliance. Orchestrating an ecosystem of high-energy assets requires a continuous integrity-assurance protocol, from climate-stabilized storage environments utilizing advanced inert-matrix suppression technologies to secure transit frameworks. If a thermal runaway event occurs during transit or storage without an audited, transparent compliance matrix, the liability instantly penetrates the operating entity, exposing the parent organization to severe financial damage. By integrating strict risk-mitigation protocols into the logistical framework, the AgTech commander ensures that operational risk is isolated. Compliance with DOT and EPA standards ceases to be a bureaucratic burden; it becomes a strategic financial fortress that secures market dominance and safeguards the enterprise's long-term capital structure.

The operational sovereignty of a high-performance AgTech fleet is intrinsically linked to the management of thermodynamic instability. Lithium, unlike other combustible materials such as hydrocarbons, does not require an external oxygen source to sustain violent combustion; its molecular architecture allows it to generate its own oxidizer during the structural collapse of the cells. For the "Owner," storing these assets is not just an inventory cost, but a high-complexity risk management task that can determine the survival of the LLC. On global fulfillment platforms like Amazon FBA, storage costs for products classified as Hazmat (Hazardous Materials) can be up to 40% higher due to facility insurance premiums, industrial zoning requirements, and the need for specialized suppression vaults. Ignoring this financial reality when projecting Total Cost of Ownership (TCO) is a management error that can erode profit margins before the first autonomous system is deployed in the field, transforming a productive asset into a latent legal liability.

From a Forensic Chemistry perspective, the phenomenon known as Thermal Runaway is an exothermic chain reaction that occurs when the internal temperature of the LiPo cell exceeds the stability threshold of the polymer separator. At a molecular level, this process begins with the decomposition of the Solid Electrolyte Interphase (SEI) layer on the anode, followed by a violent oxidation of the organic electrolyte. What makes lithium "doctorally dangerous" is the degradation of the cathode (typically lithium cobalt oxide or Nickel-Manganese-Cobalt), which releases pure molecular oxygen into the cell's internal environment. This oxygen instantly fuels the combustion of graphite, creating a thermal feedback loop that can reach 1,200°C in milliseconds. This "self-oxygenating" capability makes CO2 extinguishers useless for the fire's core; the CO2 only cools the surface, but the internal chemical reaction continues unabated, generating its own air to burn until the material is completely exhausted.

To put this risk into perspective, the Aviation Director must analyze the scale of thermal decisiveity. An average cell phone possesses about 15 Wh of energy; in case of ignition, it reaches 400°C in 3 seconds, with a heat radius of barely 12 inches. A laptop (60-90 Wh) can hit 700°C in 5 seconds, affecting an entire desk. However, a heavy-lift spray drone utilizes packs exceeding 1,500 Wh per unit; when such a pack fails, temperatures skyrocket to 900°C-1,000°C in less than 10 seconds, with a thermal flame ejection that can reach 10 feet in length. Comparatively, an Electric Vehicle (Tesla) with a 75 kWh battery represents a massive thermal load of 1,200°C that can burn for 24 hours and requires up to 26,000 gallons of water just for external cooling. The "Owner" must understand that having 10 drone batteries in a warehouse is equivalent, in terms of concentrated risk density, to having an electric vehicle fragmented into highly unstable pieces ready for a chain reaction.

The critical nature of this challenge is underscored by recent industry-wide high-energy storage thermal events, where secondary ignition cycles persisted despite initial suppression efforts, highlighting the inherent limitations of legacy cooling architectures. In the corporate sphere, a less publicized but more terrifying case involved a server room in an industrial plant in Europe, where a single defective laptop battery in a drawer triggered a thermal runaway that, failing to be contained by fire blankets, jumped to the communication racks. In less than 15 minutes, the release of hydrofluoric acid (a highly corrosive and toxic gas) optimized all the building's electronics – not by fire, but by the chemical corrosion of the lithium vapors. The loss was not just the hardware, but a total business interruption for 6 months. For an agribusiness, a similar incident in the loading warehouse during harvest would mean immediate bankruptcy.

Warehouse safety architecture must begin with a multispectral early detection system. Sensors should not be limited to detecting visible smoke but must be calibrated to identify precursor gases such as Hydrogen (H2), Carbon Monoxide (CO), and Hydrogen Fluoride (HF). These gases are released minutes before visible ignition occurs, allowing for a critical intervention window to evacuate or activate suppression systems. Ventilation is another fundamental pillar: a professional lithium warehouse requires a High-Capacity Forced Ventilation System that guarantees at least 12 to 15 Air Changes per Hour (ACH). The airflow must be designed to extract heavy gases from the floor level and light gases from the ceiling, avoiding the formation of explosive gas pockets under ATEX regulations. The floor, for its part, should be concrete with an Anti-Static Epoxy (ESD) coating to prevent a simple friction or static spark from initiating the collapse of a degraded cell.

Regarding fire suppression, equipment distribution must follow NFPA 855 standards. For industrial storage, one Liquid Encapsulating Agent (such as F-500) or specialized battery-rated suppression equipment is recommended for every 50 square meters, ensuring a response time of seconds. While CO2 is effective for protecting charger electronics without leaving residue, the battery fire itself requires a liquid agent that can penetrate the casing and cool the core below the ignition temperature. Furthermore, the use of Pallet Containment Blankets made of calcium silicate fibers is mandatory to isolate incidents. These blankets act as a thermal sarcophagus: if one pallet ignites, the blanket prevents the radiant heat (which expands in a thermal bell shape) from reaching adjacent packs, limiting the disaster to a single unit rather than the entire fleet.

Personnel training must be at the Hazwoper level. An operator who does not know the principles of lithium is a risk to the LLC. Training must be rigorous: learning to identify the "sweet smell" of leaked electrolyte, the hiss prior to explosion, and the handling of expanded vermiculite kits. Training must include drills where it is understood that water, in contact with burning lithium, can generate hydrogen explosions. This safety culture elevates the LLC's prestige during audits by firms like PwC or Deloitte and ensures that human capital is shielded against accidents that occur due to technical complacency. Safety is not a manual on a shelf; it is a daily practice of visual inspection and voltage monitoring that the "Owner" must personally supervise as part of their operational sovereignty.

Regulatory compliance is the documentary trail that protects the Owner's Corporate Veil. Organizations must align with ISO 14001 (Environmental Management) and ISO 45001 (Occupational Health and Safety), but it is ISO 9001 that guarantees traceability: ensuring absolute granular visibility over the chemical provenance and manufacturing lineage of every high-energy asset within the inventory.These leading companies invest billions in pressure relief systems (Vents) that safely channel gases, but none of this matters if the "Owner" stores batteries on conventional wooden racks. Before establishing a warehouse, it is vital to conduct a safety audit validating separation distances of at least 1.5 meters between battery groups, minimizing the cumulative fuel load and ensuring that drainage systems do not compromise the integrity of the local water table in the event of an emergency containment failure.

Finally, the "Owner" must understand that the lithium culture is a culture of Continuous Vigilance. Resilience is not bought with extinguishers; it is built with materials engineering and telemetry monitoring protocols that alert to temperature increases in the warehouse overnight. By integrating these safety standards and advanced chemical knowledge into the LLC's DNA, the Owner transforms a "potential time bomb" into an engine of technological sovereignty. Mastery over lithium is what allows a fleet to be truly autonomous and safe, ensuring that the future of precision agriculture is as stable as the ground it flies over and as professional as the mind that directs it, ensuring your investment remains protected under the strictest standards of the modern aerospace industry.

CHAPTER 16

THERMAL RUNAWAY

The Thermal Time Bomb: Anatomy of a Lithium Fire To master the future of AgTech, we must confront the industry's most silenced truth: the inherent risk of high-density energy. Industrial drones are powered by massive blocks of Lithium-Polymer (LiPo) or Lithium-Ion batteries.

When one of these aircraft suffers an impact (due to pilot error, material

fatigue, or a bird strike), or when batteries are overexerted under the extreme heat of the agricultural summer, the delicate separator membrane between the anode and cathode inside the cell can fracture.

This triggers the phenomenon known as Thermal Runaway. The Occupational Safety and Health Administration (OSHA) and the National Fire Protection Association (NFPA) warn that lithium fires do not behave like conventional fires. During Thermal Runaway, the chemical degradation of the battery generates its own oxygen. The temperature escalates from 20°C to over 1,000°C (1,832°F) in a matter of seconds, expelling toxic gases and flaming shrapnel. The fire is self-sustaining. It does not need outside air to burn, making traditional extinguishing methods a deadly illusion.

16.1 Active Suppression

Every field support vehicle must be equipped with AVD Lith-Ex extinguishers (6 to 9 liters) certified under EN3-7 and NTA 8133. When discharged, the vermiculite seals the battery in an incombustible mineral film, isolating the cell, cooling the core, and blocking oxygen transfer without damaging the soil ecosystem. For massive charging stations in the farm's hangars, heavy equipment such as 25 to 50-liter AVD mobile units (Trolleys) is required.

16.2 Thermal Shielding

If the fire cannot be attacked in its first few seconds, the priority shifts to isolating the extreme thermal radiation to protect other aircraft, vehicles, and the dry harvest itself. This is where industrial-grade fire blankets certified under DIN SPEC 91489:2024-11 come into play. These are not ordinary fabrics; they are constructed from oxidized glass textiles and coated with graphite and silicone compounds. They withstand continuous temperatures of 1,000°C (with proven peaks up to 1,100°C). Deploying a 6x8 meter blanket over a crashed drone and its payload instantly confines the threat.

CHAPTER 17

CHEMICAL LIABILITY (PFAS)

17.1 EPA Risks

The Fatal Reflex When an industrial drone carrying high-density lithium batteries crashes in a field and initiates thermal runaway, the ground crew's immediate instinct is to grab the nearest fire extinguisher. In most agricultural operations, this is a standard ABC dry chemical or an AFFF (Aqueous Film Forming Foam) extinguisher purchased from a local hardware store. For the modern farm manager, this single, panicked decision can cause exponentially more financial ruin than the fire itself. You may succeed in extinguishing the flames surrounding the drone, but in doing so, you have just introduced a catastrophic chemical liability into your primary asset: your soil. The Anatomy of "Forever Chemicals" to understand the risk, we must look at the chemical composition of traditional fire suppression agents. Many low-cost extinguishing foams rely heavily on PFAS (Per- and Polyfluoroalkyl Substances). These synthetic compounds are highly effective at smothering conventional fires by creating a film over the fuel source. However, PFAS are universally known in the scientific and environmental communities as "forever chemicals."

The carbon-fluorine bond in PFAS is one of the strongest in nature. Once deployed in an agricultural field, these chemicals do not break down under sunlight, water, or microbial action. They seep into the topsoil, leach into the groundwater, and are eventually absorbed by the root systems of the crops you are trying to protect.

The Regulatory Guillotine and the Death of Organic Certification The agribusiness sector is currently facing a massive regulatory crackdown. Agencies such as the Environmental Protection Agency

(EPA) in the United States, and equivalent bodies in the European Union, are enacting aggressive legislation to track, penalize, and ban PFAS contamination.

17.2 Organic Certifications

If an operational manager utilizes a PFAS-laden suppression agent to address a thermal runaway event in high-value specialty assets – such as a certified organic orchard or a premium produce plantation – the consequences are swift and irreversible:

1. Loss of Premium Certifications: The USDA Organic certification, Global G.A.P., and strict export licenses have a zero-tolerance policy for PFAS contamination. A single deployment of toxic foam can result in the immediate revocation of these certifications for the affected acreage.
2. Export Rejection: High-yield agricultural businesses rely on international markets. Crops testing positive for PFAS absorption face immediate rejection at customs, particularly in European and Asian markets.
3. Environmental Remediation Costs: The cost of excavating, treating, and replacing PFAS-contaminated soil far exceeds the cost of a commercial drone. The False Economy of Cheap Suppression.

This brings us to the ultimate paradox of agricultural drone safety. A commercial agricultural drone, fully equipped, is a depreciating asset valued between $20,000 and $40,000. It is replaceable, and if properly managed, it is insured. Your land, however, is a permanent, appreciating asset worth millions.

CHAPTER 18

INDUSTRIAL CONTAINMENT

18.1 The NTA 8133 Standard

The industrial management of lithium-ion risks requires a shift from general fire safety to battery-specific certification. The NTA 8133 represents the first dedicated standard for testing extinguishing agents against lithium battery fires. Under this protocol, the use of Aqueous Vermiculite Dispersion (AVD) is validated as the only mineral-based solution capable of creating an oxygen-proof film that encapsulates the thermal runaway process. For the enterprise, adopting NTA 8133-certified equipment is the primary technical defense against operational paralysis and the first step toward achieving industrial-grade safety compliance in agricultural fleets.

Industrial containment is the absolute baseline of operational sovereignty. Operating a high-density aerospace fleet without a mathematically rigid containment architecture is not an oversight; it is an uninsurable liability. Thermal runaway within high-capacity lithium-ion storage is not a hypothetical risk but a statistical certainty inherent to the electrochemistry of aggressive charge cycles. The NTA 8133 standard, developed as the definitive European benchmark for the safe storage of lithium-ion cells, serves as the uncompromising foundation for our containment doctrine. Standard safety protocols mandated by the Occupational Safety and Health Administration (OSHA) often fail to account for the unique, self-sustaining oxygen generation of a lithium fire. By actively integrating the NTA 8133 framework into the operational matrix, the AgTech commander establishes compartmented fire resistance, rigorous ventilation physics, and automated detection thresholds. This is not merely about physical safety. Aligning the enterprise's energy storage protocols with elite global standards instantly collapses insurance premiums, shields the corporate veil from negligence liability, and transforms the autonomous operational unit into a fortress of audited compliance.

18.2 DIN SPEC 91489 Blankets

When a tier-one autonomous aerospace battery enters thermal runaway, it breaches 1,000°C in a matter of seconds, releasing toxic fluoride gases and overwhelming standard suppression systems. Deploying water or conventional Class ABC chemical extinguishers on a lithium-ion fire is mathematically and chemically useless, often exacerbating the exothermic reaction. The enterprise must deploy specialized isolation infrastructure governed by the strict parameters of DIN SPEC 91489. This specification dictates the material science required for active lithium fire containment, demanding the deployment of high-temperature suppression blankets engineered with Aqueous Vermiculite Dispersion (AVD) or specialized woven silica. These advanced materials physically isolate the thermal event, starving the perimeter of collateral fuel and absorbing the extreme thermal shock. Under The Physics of Capital™, the DIN SPEC 91489 blanket is not viewed as emergency equipment; it is a financial tourniquet. By isolating a localized thermal event within seconds, the blanket prevents the catastrophic contagion from optimizing the entire logistics vehicle, thereby rescuing hundreds of thousands of dollars in adjacent capital assets from instantaneous vaporization.

18.3 FRC Quarantine Operations

The logistical management of a compromised battery does not end with the extinction of flames. Fire Resistant Containers (FRC) and dedicated quarantine protocols are essential for the post-event phase. A battery that has experienced thermal stress remains a volatile chemical liability for at least 72 hours. Standardizing quarantine operations ensures that damaged units are isolated in specialized enclosures that prevent reignition or toxic gas leakage. This logistical layer is what separates a professional AgTech operation from a high-risk pilot activity, ensuring that a single unit failure does not escalate into a total facility loss.

Containment does not end with suppression; it transitions immediately into quarantine logistics. Once a thermal anomaly is isolated, the compromised lithium asset instantly triggers severe federal regulatory oversight. Under Title 49 of the United States Code of Federal Regulations (49 CFR), a damaged or defective lithium-ion battery is classified as a high-risk Hazardous Material. Moving this compromised asset without strict adherence to the Pipeline and Hazardous Materials Safety Administration (PHMSA) mandates invites devastating federal audits and catastrophic civil penalties. The execution of Fire Resistant Container (FRC) Quarantine Operations bridges the gap between field failure and legal compliance. The compromised cell must be immediately sequestered within a certified FRC, engineered with specialized filtration to manage toxic off-gassing and prevent secondary ignition during transit. This rigid quarantine protocol ensures that a localized hardware failure never escalates into a systemic corporate liability. Operating an elite AgTech fleet requires understanding that mastering the physics of failure is just as critical as mastering the physics of flight.

CHAPTER 19
THE INSURANCE CRISIS

19.1 Corporate Safety Protocols

The global insurance architecture is currently undergoing a systemic crisis of ignorance. Underwriters and actuaries are attempting to price the risk of advanced aerospace infrastructure using actuarial tables built for manned, diesel-powered terrestrial machinery and hobbyist drones. This is a mathematical atrocity. When a traditional agricultural operation applies for commercial liability coverage, the insurer assumes a baseline of inevitable human error, fatigue, and mechanical degradation. The National Transportation Safety Board (NTSB) consistently attributes over 80% of aviation and heavy machinery accidents to operator cognitive failure, fatigue, or spatial disorientation. Legacy safety protocols are inherently flawed because they attempt to regulate the biological operator through bureaucratic paperwork and high-visibility vests. Under the doctrine of The Physics of Capital™, we do not attempt to regulate human friction; we amputate it entirely from the operational matrix.

Corporate safety is no longer a human resources exercise; it is an algorithmic absolute. True safety protocols are written in code, embedded in the edge-computing architecture of the autonomous swarm, and executed without emotional or physical deviation. To satisfy and overwhelm the strict mandates of the Federal Aviation Administration (FAA) Part 137 commercial operations, the AgTech sovereign must construct an impenetrable digital perimeter. This involves deploying geofencing algorithms with millimeter precision, redundant obstacle-avoidance telemetry, and automated Return-to-Launch (RTL) fail-safes triggered by minute drops in electrochemical voltage. Furthermore, the Occupational Safety and

Health Administration (OSHA) requires rigorous containment of hazardous chemicals. By executing Ultra-Low Volume (ULV) payloads via autonomous vectors, the enterprise physically removes the human worker from the toxic exposure zone, instantly collapsing the workers' compensation liability profile. When safety is digitized, it ceases to be a variable cost and becomes a securitized corporate asset. The operational logs generated by the swarm do not merely record a flight; they mathematically prove the absence of negligence. This audited, immutable record protects the corporate veil with a ferocity that no traditional safety manual could ever achieve.

19.2 Dictating Premium Costs

The fatal mistake made by nascent AgTech operators is treating insurance premiums as a fixed, non-negotiable tax imposed by the market. Requesting a quote from an underwriter is a submissive posture that immediately surrenders financial sovereignty. Insurance brokers thrive in environments of informational asymmetry; they charge astronomical premiums when they cannot mathematically comprehend the operational risk. The commander operating under The Physics of Capital™ does not ask for an insurance rate. We dictate the premium cost through the aggressive weaponization of operational telemetry.

Actuarial science is entirely dependent on historical loss ratios. Because the insurance industry lacks a decade of loss data for high-density, AI-governed aerospace swarms, they default to maximum-risk pricing models. To break this actuarial siege, the enterprise must present a forensic, data-driven mandate to the underwriter. By aggregating flight logs, predictive maintenance cycles, thermal management records, and verifiable chemical containment reports, the operator transitions the conversation from speculative risk to deterministic physics. The Insurance Services Office (ISO) and elite underwriting syndicates rely on verifiable risk-mitigation data to construct their models. When we deliver a comprehensive matrix proving that our fleet utilizes algorithmic collision avoidance, isolated lithium containment architectures (such as the NTA 8133 standard), and real-time kinetic monitoring, we systematically dismantle their justification for high premiums.

This is financial engineering at its most ruthless. The Securities and Exchange Commission (SEC) increasingly mandates the disclosure of material risks, including operational and environmental liabilities, for securitized agricultural portfolios. By leveraging autonomous infrastructure to drive the probability of an environmental spill or a catastrophic kinetic impact mathematically close to zero, the enterprise achieves a massive reduction in the cost of capital. We force the insurance syndicate to recognize that insuring our fleet is not a gamble, but a mathematical certainty of zero-claim operations. The premium collapses not because we negotiated well, but because the physical risk no longer exists. Dictating the cost of indemnity is the ultimate leverage; it transforms insurance from a margin-crushing expense into a strategic barrier to entry that our under-capitalized competitors cannot survive.

PART V

SCALING, LOGISTICS & DEFENSE

CHAPTER 20

THE PAYLOAD PARADOX

20.1 Arms Race of Volume

The contemporary AgTech sector is paralyzed by an irrational arms race dictated by marketing departments rather than aerospace engineers. Historically, manufacturers introduced baseline capacities of 30, 40, and 50 liters as evolutionary steps to test material fatigue and flight algorithm stability. Instead of optimizing these agile platforms, the industry blindly leaped toward 100-liter leviathans. They operate under the archaic premise that moving more liquid inherently equates to higher productivity. This is a dangerous agronomic fallacy that ignores the specific gravity of the inputs. In precision agriculture, operators do not spray pure water; they deploy dense chemical suspensions, emulsions, and biosolutions that are significantly heavier than water per unit of volume. A 100-liter tank filled with a dense chemical suspension acts as a massive static payload, not counting the chassis and the battery arrays. As this mass increases, the kinetic energy required to maintain aerodynamic lift forces exponential amperage consumption, actively optimizing the lithium-ion matrix. By prioritizing raw volumetric capacity over fluid thermodynamics, the manufacturing industry is not innovating; it is merely building inefficient aerial tractors. Absolute hegemony requires the systemic elimination of volume through cognitive precision.

20.2 Logistical Footprint (20L to 60L)

The logistical footprint is the absolute dictator of the Total Cost of Ownership (TCO). Transitioning from agile 20-liter to 60-liter platforms toward the gigantism of 100-liter architectures introduces a paralyzing systemic friction. A forensic audit comparing an optimized mid-capacity paradigm, specifically the highly engineered 70-liter frame, against an inefficient 100-liter model optimizes the commercial myth of heavy lifting. The superiority of the optimized mid-capacity paradigm is proven through the principles of aerodynamic drag. When a massive 100-liter vector faces crosswinds, its enormous frontal area acts as a parachute. Just as an automobile loses significant engine power operating at high altitudes due to oxygen density, a 100-liter drone exposed to standard field winds suffers a drastic drop in efficiency. The motors must absorb massive current spikes simply to correct the drift generated by its own gigantic fuselage. As the foundational laws of electrical heating state, the heat generated inside a battery is proportional to the square of the electrical current. This massive amperage literally cooks the cells from the inside out. In stark contrast, an aerodynamically optimized 70-liter vector slices through wind resistance with marginal power consumption, stabilizing the flight and protecting the lifespan of the electrical capital. Structural gigantism is thermodynamic suicide against the elements.

20.3 The Swarm Fallacy

There is a corporate illusion suggesting that deploying multiple heavy drones will exponentially multiply crop coverage. We must question if adding more dead weight actually solves a logistical bottleneck. This fallacy ignores basic kinematics. Productivity is strictly a function of distance covered over time. It is not about the sheer volume of liquid that leaves the ground. A 100-liter drone is confined to slow operational speeds due to high gravitational inertia and wind resistance. The shifting of a large liquid payload destabilizes the gyroscope if the unit attempts to accelerate or execute sharp turns at the edge of a field. This risks a total system failure. An optimized mid-capacity vector dominates field kinematics. It maintains tactical speeds and agile turning capabilities. It covers significantly more linear distance per hour than the heavier model. The mathematical limit for optimal swarm saturation is strict. A human operator can manage a maximum of three to four optimized vectors. Deploying five or more units instantly creates a thermal bottleneck in the terrestrial microgrid. Attempting a swarm with large batteries overloads the ground generators and grounds the entire fleet. We must protect the operational expenditure and the daily cash flow. Hegemony is achieved through continuous staggered rotation and speed. It is never achieved through heavy stagnation.

20.4 Administrative Burden

Aerospace gigantism attracts relentless federal scrutiny. In the United States airspace, transitioning from agile platforms toward heavy-lift 100-liter models violently breaches the standard commercial operational thresholds set by the Federal Aviation Administration (FAA). Scaling up payload volume rapidly pushes the Maximum Takeoff Weight (MTOW) into categories that demand complex exemptions for heavy aircraft. The industry naively assumes that securing these federal exemptions is a simple administrative task; the reality is a capital trap. This level of tonnage attracts near-military jurisdiction, forcing the enterprise to deploy additional Visual Observers (VO), certified predictive maintenance personnel, and strict kinetic impact mitigation protocols under the Department of Transportation (DOT). Any projected marginal revenue from spraying a larger volume is immediately devoured by compliance fees, redundant personnel salaries, and astronomical liability insurance premiums dictated by the kinetic risk of a heavy vehicle operating at low altitudes.

20.5 Unresolved Horizons

The AgTech sector stands at a critical evolutionary vertex. The future of agricultural sovereignty will not be conquered by engineers who simply mold larger plastic tanks to systemic optimization the wind with brute force. Perpetual disruption demands the complete eradication of massive liquid dependency through chemical and algorithmic evolution. The unresolved horizon lies in the manipulation of fluid rheology – adapting the viscosity and surface tension of highly concentrated chemicals so that a microscopic payload can withstand wind currents without suffering aerodynamic drift. The ultimate objective is not to figure out how to lift 100 liters with greater battery efficiency; it is to re-engineer the biological intervention so that a tactical 10-liter dose achieves total pathogen systemic elimination, executed by a kinematically perfect swarm. Until the industry wakes up from the fallacy of volume and embraces the physics of pure thermodynamic precision, the absolute financial potential of the autonomous matrix will remain an unresolved horizon.

CHAPTER 21
TECHNOLOGICAL ENDOGENEITY

21.1 Vendor Lock-in Trap

Technological endogeneity is the silent cancer of the autonomous sector. Monolithic manufacturers design their architectures not to protect the biological asset, but to hijack the operator's cash flow in perpetuity through closed ecosystems. Surrendering flight matrix control to a vendor that encrypts the thermodynamics of its motors or its battery health is the exact definition of corporate negligence. Relying blindly on a foreign "black box" constitutes a breach of fiduciary duty to investors.

The operational cost of this submission is brutal. A closed-ecosystem agricultural drone forces the owner into artificially inflated predictive maintenance costs, averaging between $2,500 and $4,500 USD annually per unit, demanding the equipment be shipped back to authorized service centers. This generates a two- to four-week operational paralysis during peak spraying season. Under the doctrine of The Physics of Capital™, the machine must be agnostic and modular. True resilience demands that any critical component be field-swappable by the operator in under 45 minutes, utilizing standardized industrial-grade parts. You are a financial hostage if your flight capacity depends on an international vendor shipping a coded replacement part.

21.2 Corporate Intelligence Shield

The aerospace chassis is irrelevant plastic. The true asset is the information. The agricultural sector has been classified by the

FBI and the Department of Homeland Security (DHS) as highly vulnerable critical infrastructure. Cybersecurity has bypassed the phase of protecting passwords; we have entered the era of cyber-kinetic threats, where an intrusion into the swarm can alter the Variable Rate Application (VRA) dosing to deliberately scorch hundreds of acres.

The Corporate Intelligence Shield demands military-grade protocols. Security updates cannot rely on intermittent USB connections or monthly schedules. The infrastructure must execute end-to-end encrypted Over-The-Air (OTA) patches, processing malware signatures in milliseconds through Edge Computing architectures. Adhering to the National Institute of Standards and Technology (NIST 800-53) frameworks, the swarm must operate under a strict Zero Trust model. If the drone manufacturer retains remote access (backdoors) to your telemetry to "improve service," they virtually own your agronomic intelligence and can package it to sell to speculative hedge funds.

21.3 Logistical Collapse

Biology does not negotiate with server outages. Modern autonomous logistics no longer depend exclusively on hardware failures. The most decisive vulnerability today is reliance on third-party software and cloud computing. When an operator anchors their flight matrix to the manufacturer's remote cloud, they expose the enterprise to algorithmic collapse.

To permanently eradicate remote server dependency, the AgTech sovereign must deploy a localized, offline-first network architecture. The mobile command center must act as a self-contained local area network (LAN), utilizing decentralized Edge Servers and independent Real-Time Kinematic (RTK) base stations that do not require cellular LTE/5G triangulation. The swarm communicates through encrypted radio frequency mesh networks, processing topography and obstacle avoidance locally. The most frequent failures that annihilate profitability do not occur in the field; they occur in remote data centers. Application Programming Interface (API) interruptions, latency drops in services like Amazon Web Services (AWS), or forced, defective firmware updates can electronically "brick" an entire fleet in seconds. True logistical sovereignty lies in severing this digital umbilical cord. If the swarm cannot take off, map, and spray completely off-grid, the enterprise is operating under an illusion of control.

21.4 Cost of FAA Compliance

Navigating commercial airspace demands absolute transparency and liquid capital. The industry sells the lie that securing Federal Aviation Administration (FAA) certification is a minor administrative task. The quantitative reality is punitive. Securing a Part 137 and Section 44807 exemption to operate heavy-lift aircraft requires forensic audits of flight manuals. Corporate aviation law firms in the United States demand retainers ranging from $8,000 to $15,000 USD simply to structure and file exemption petitions for heavy swarms.

Added to this initial CAPEX is the operational expenditure (OPEX) of underwriting. Insurance syndicates impose civil liability premiums starting at $3,000 to $7,000 USD annually per heavy drone – rates that fluctuate violently if the hardware belongs to a closed ecosystem that prohibits black-box extraction after a crash. If a collision occurs and the National Transportation Safety Board (NTSB) cannot audit the drone's source code to determine liability, the corporate veil is instantly pierced, and executives bear the total civil cost. Regulatory compliance is not a flat fee; it is a relentless toll for those who operate opaque technology.

CHAPTER 22

THE AVIATION DIRECTOR

The Paradigm Shift in Agricultural Management: From Farmer to Aviation Director The transition from manual agricultural labor to autonomous aerial systems is no longer a futuristic concept; it is the cornerstone of operational survival. As we have explored throughout this guide, labor shortages and the escalating costs of temporary work visa programs have positioned high-capacity commercial drones as the primary asset for reducing the Total Cost of Ownership (TCO).

However, the greatest mistake made by modern agribusiness is the oversimplification of risk. Purchasing an industrial heavy-lift agricultural drone is not acquiring a faster tractor; it is inaugurating a corporate aviation department. The Federal Aviation Administration (FAA) no longer views these machines as experimental farm tools, but as heavy commercial aircraft operating in national airspace. The modern farm manager must adopt an aerospace mindset, where logistical foresight, catastrophic risk mitigation, and corporate liability dictate financial success.

The Swarm Illusion and Airspace Physics There is a dangerous narrative pushed by technology marketing: the myth of absolute automation through drone "swarms." Dazzled by nighttime LED drone shows, many farm owners believe they can buy multiple heavy-capacity drones and fly them simultaneously over their crops with the push of a single button.

Aerodynamic reality is unforgiving. A fully loaded commercial drone generates massive wake turbulence. Attempting to operate in dense formations over a high-yield crop without advanced deconfliction (collision avoidance) protocols invites disaster. Furthermore, saturating the radio frequency (RF) spectrum by transmitting heavy telemetry from multiple aircraft simultaneously elevates the risk of a "Fly-Away" (loss of command link) event. Responsible scaling does not mean filling the sky with drones; it means establishing a system of continuous relays.

22.1 Human Capital

The traditional agricultural sector suffers from a chronic atrophy in its understanding of human capital. Legacy operators continue to make the catastrophic financial error of hiring manual laborers, handing them a remote control, and expecting aerospace superiority to somehow materialize. This is a primitive and unacceptable allocation of corporate risk. Under the doctrine of The Physics of Capital™, an Aviation Director is not a manual pilot; manual piloting of an autonomous vector is an obsolete technical redundancy in the era of artificial intelligence and RTK kinematic navigation. This individual is the central cognitive node of a high-density logistical matrix. The United States Department of Labor (DOL) annually reports paralyzing inflation and critical shortages within the H-2A agricultural visa program, rendering reliance on manual labor an unsustainable operational liability. The transition toward an autonomous swarm does not merely replace the worker; it elevates the human function from physical labor to aerospace command.

Human capital in an elite AgTech operation is strictly measured by the capacity to process thermal, electrical, and chemical saturation in real-time. When a fleet of optimized 70-liter class vectors is airborne executing Ultra-Low Volume (ULV) applications, the director is not watching the hardware; they are auditing the telemetry. Their function is to anticipate the thermodynamic stress of the charging micro-grids, orchestrate the quarantine of lithium cells exhibiting State of Health (SoH) voltage fluctuations, and synchronize flight paths to prevent aerodynamic downwash interference. Relegating this professional to the status of a "crop duster" ignores the reality that total responsibility for hundreds of thousands of dollars in assets and the biosecurity of a multi-million-dollar crop rests entirely in their hands. Demanding anything less than military discipline and absolute financial acuity in this position guarantees the evaporation of the EBITDA margin.

22.2 The Paper Trail

Federal regulatory agencies do not exclusively penalize kinetic accidents or chemical spills; they penalize the absence of forensic documentation with implacable fury. The industry sells the fantasy that autonomous flight is free of bureaucracy, but operational reality dictates that the paper trail is the only fortress separating the enterprise from legal systemic elimination. Operating autonomous vectors that deploy liquid chemical payloads under Federal Aviation Administration (FAA) Part 137 exemptions instantly transforms an agricultural field into a highly regulated commercial airspace. The FAA, the Environmental Protection Agency (EPA), and the Department of Transportation (DOT) have absolutely no interest in the biological yield of the crop; they demand immutable, encrypted, and auditable proof of continuous regulatory compliance.

Maintaining fragmented logbooks on paper or unencrypted spreadsheets is an open invitation to catastrophic civil liability. Every battery charge cycle, every Over-The-Air (OTA) firmware update, every predictive maintenance check of the rotors, and every ounce of decisive active ingredient transported must be recorded in a decentralized database. If a vector experiences a redundancy failure and crashes into civil infrastructure, the National Transportation Safety Board (NTSB) will confiscate the operational logs before the smoke clears. If the Aviation Director can present an unbreakable digital trail proving that the algorithm failed despite 100% logistical and maintenance compliance, the corporate veil remains intact, and the liability is transferred to the hardware manufacturer or the insurance syndicate. The paper trail is not an administrative burden for the commander; it is the kinetic shield that protects the board of directors and insulates capital against corporate negligence lawsuits.

22.3 Forensic Auditing

Financial hegemony is not a biological accident; it is the mathematical result of ruthlessly auditing operational entropy. Forensic auditing is the corporate transition from absorbing expenses to dictating margins. The Aviation Director must operate simultaneously as an actuary and a thermodynamic engineer. By crossing kinematic flight logs with Variable Rate Application (VRA) maps, the commander extracts the exact Total Cost of Ownership (TCO) per sprayed acre in real-time. This forensic rigor detects the capital leakage that remains entirely invisible to the traditional operator. If telemetry reveals that a vector is consuming 4% more amperage during yaw maneuvers due to micro-abrasions on the propellers, the forensic auditor orders the immediate replacement of the rotors. To the naked eye, the hardware was functioning perfectly, but the mathematics proved that the 4% resistance was internally cooking the battery matrix via the Joule effect, threatening to optimize a high-value asset thousands of cycles before its scheduled depreciation.

This hyper-vigilance directly connects the field to the offices of the Securities and Exchange Commission (SEC) for corporate operations and the Internal Revenue Service (IRS). A perfectly audited forensic history allows the organization to aggressively maximize depreciation advantages under the Modified Accelerated Cost Recovery System (MACRS). It proves exactly when an electrical asset has ceased to be efficient, justifying its immediate replacement through capital subsidies or tax deductions. The drone ceases to be an agricultural tool and reveals its true nature: a rotary engine for asset securitization. Forensic auditing guarantees that every drop of carrier liquid, every injected electron, and every millisecond of flight is strictly subordinated to the violent expansion of corporate capital.

CHAPTER 23

AUTOMATING RESILIENCE

23.1 Damage Assessment

Resilience is a mathematical equation, not an emotional response. When an autonomous aerospace vector suffers a catastrophic kinetic failure, whether induced by unpredictable atmospheric micro-bursts or a sudden electrochemical voltage collapse, the legacy agricultural response is operational paralysis. The traditional operator walks into the field, stares at the shattered carbon fiber, and begins a slow, deeply inaccurate process of manual inspection. This is an unacceptable hemorrhage of time. A human being cannot forensically diagnose a breached Solid Electrolyte Interphase (SEI) layer, nor can they reconstruct the aerodynamic yaw failure that precipitated the crash. Under the doctrine of The Physics of Capital™, damage assessment is executed in the milliseconds prior to impact. The autonomous matrix utilizes its Edge Computing architecture to instantly lock and encrypt the final sequence of telemetry, cross-referencing G-force anomalies with thermal escalation.

The Federal Aviation Administration (FAA) and the National Transportation Safety Board (NTSB) do not accept human testimony in the aftermath of an incident involving heavy-lift commercial aircraft; they demand raw, unalterable data. By automating the forensic capture of a crash, the system isolates the exact point of failure. It proves definitively whether the event was an unavoidable Act of God, a third-party software glitch originating from a remote server, or a manufacturer hardware defect. The crash is violently stripped of its chaos. It is instantly transformed into a structured, auditable dataset. This automated assessment legally shields the Aviation Director from accusations of negligence, ensuring that the corporate veil remains impenetrable even in the face of catastrophic hardware loss.

23.2 Instant Insurance Claims

The global insurance industry is built on a foundation of actuarial delay. Underwriters leverage time, utilizing archaic claims adjustment processes to hold corporate capital hostage while they look for loopholes to deny coverage. When an operator relies on manual claim filing, the enterprise bleeds EBITDA every single hour a replacement vector is not in the air executing Ultra-Low Volume (ULV) biological interventions. The AgTech sovereign annihilates this bureaucratic friction through the aggressive deployment of parametric insurance protocols and automated telemetric reporting. The moment the decentralized command center registers a kinetic impact exceeding the structural threshold of the chassis, the system bypasses human intervention entirely.

Before the wreckage has even stopped smoldering, the localized network compiles the encrypted flight logs, verified maintenance records, and real-time topographical weather data, packaging them into an immutable forensic dossier. This data is transmitted directly via API to the insurance syndicate. There is no room for an adjuster's subjective interpretation. When the data proves mathematically that the vector was operating strictly within the legal parameters of its Part 137 exemption and suffered an unrecoverable failure, the parametric trigger forces an immediate financial settlement. We do not negotiate with adjusters; we overwhelm them with forensic truth. This ruthless automation secures immediate liquidity. The optimized asset is financially resurrected and replaced before the biological window of the crop closes. Resilience is not bouncing back; it is the instantaneous weaponization of data to guarantee the uninterrupted, violent expansion of capital.

CHAPTER 24

FREE CAPITAL & SUBSIDIES

24.1 Navigating USDA EQIP

Non-repayable federal capital is not a subsidy; it is a strategic transfer of sovereign liquidity designed to absorb operational risk for elite technology platforms. The Environmental Quality Incentives Program (EQIP), orchestrated by the USDA Natural Resources Conservation Service (NRCS), is routinely mismanaged by legacy agricultural operators who approach the agency with a submissive posture, begging for minor cost-share allocations to fund heavy, soil-compacting terrestrial machinery. This is a profound miscalculation. Under the strict framework of The Physics of Capital™, federal conservation mandates are a mechanism for institutional leverage. The state is legally compelled to distribute capital to operations that demonstrably mitigate systemic environmental risks.

By deploying decentralized, autonomous aerospace infrastructure executing Ultra-Low Volume (ULV) liquid applications, the enterprise aligns perfectly with NRCS Conservation Practice Standard 595 for Integrated Pest Management. The data-driven repeatability of this efficiency allows the operator to present an unassailable forensic case to federal auditors. We prove the absolute eradication of chemical runoff, the total elimination of off-target drift, and the preservation of soil macroporosity by permanently removing heavy steel axles from the field. This level of audited technical precision transforms the application process from a speculative field service into a compliance-backed conservation asset. The sovereign enterprise does not rely on government charity. We leverage the state's own environmental metrics to force the federal government to finance our aerospace infrastructure, expanding the asset base while systematically insulating corporate cash flow from initial capital expenditures.

24.2 Federal Subsidies

The fatal operational bottleneck of government subsidies is their retroactive, reimbursement-based structure. Nascent AgTech organizations routinely hemorrhage liquidity while waiting for bureaucratic clearing cycles, paralyzing their deployment schedules. To achieve absolute financial hegemony, the Aviation Director must execute a ruthless triangulation between federal reimbursement frameworks, the Farm Credit System, and the internal revenue tax code. Capital must never remain stagnant. The enterprise leverages the guaranteed EQIP allocation or federal conservation contracts as an unassailable financial guarantee to secure immediate, low-interest credit lines from institutional agricultural banks. We do not use corporate equity to purchase hardware. The bank finances the initial acquisition, the federal subsidy absorbs the operational cost via reimbursement, and the IRS Modified Accelerated Cost Recovery System (MACRS) combined with Section 179 allowances compresses the remaining asset depreciation to zero on the corporate balance sheet.

This is the ultimate manifestation of the repeatability axiom. Because our autonomous flight matrix operates with deterministic precision, the variance in chemical deployment and energy consumption approaches zero, delivering the exact risk profile required by institutional underwriting syndicates. Financial institutions and private equity funds are aggressively seeking to place capital into verified Environmental, Social, and Governance (ESG) frameworks. An autonomous, zero-emission electric swarm that collapses chemical waste by 25% is the ultimate vehicle for ESG capital allocation. By translating raw telemetric operational data into verifiable environmental credits, the enterprise dictates interest rates and gains access to near-zero-cost capital pools. The drone ceases to be an operational expense or a localized field tool; it is transformed into a high-velocity capital extraction engine that leverages federal policy to establish an unshakeable, non-replicable

CHAPTER 25

THE ROI MATRIX (B2B)

25.1 Owner-Operator vs. Contractor

The corporate boardroom is currently suffocating under a false dichotomy. When a massive agribusiness conglomerate seeks to integrate autonomous aerospace technology, the financial directors instinctively default to a binary choice: either purchase the fleet outright and assume the role of the Owner-Operator, or outsource the entire operation to a third-party Contractor. Both of these traditional pathways are fundamentally flawed. They represent an archaic misunderstanding of how capital actually behaves when subjected to the laws of physical and operational friction.

Assume the role of the Owner-Operator, and the enterprise instantly absorbs the total sum of operational entropy. The board of directors signs a capital expenditure (CAPEX) authorization, buying depreciating plastic, highly volatile lithium-ion matrices, and the staggering regulatory burden of Federal Aviation Administration (FAA) Part 137 compliance. The corporation traps its liquid capital in hardware that begins obsolete depreciation the moment it leaves the factory. This is the financial equivalent of potential energy bleeding out into the atmosphere as useless heat. Conversely, the traditional Contractor model is equally corrosive. Hiring a third-party custom applicator by the acre reduces the autonomous swarm to a cheap commodity. It introduces severe informational asymmetry. The contractor is financially incentivized to spray as fast as possible, ignoring fluid dynamics, aerodynamic downwash, and biological absorption rates just to meet a daily quota. The contractor gets paid for the motion, while the corporate landowner absorbs the biological failure.

To shatter this matrix, we must look beyond agriculture and apply the highest principles of commercial real estate development and joint-venture architecture. Consider the visionary real estate developer approaching a land-rich, cash-poor property owner. The amateur developer attempts to buy the land at a discount, triggering a hostile negotiation. The sovereign developer, however, proposes a radically different geometry: "Do not sell me the dirt. You provide the land; I will provide the capital, the architecture, and the concrete to build a premier commercial plaza. We will not act as buyer and seller; we will act as strategic partners, and we will split the ongoing commercial leases in perpetuity."

This is the exact paradigm shift required in the AgTech B2B matrix. The landowner, the corporate farm, holds the biological substrate. The autonomous operator holds the kinetic energy, the algorithmic intelligence, and the aerospace infrastructure. We do not sell them a drone. We do not charge them a miserable fee per sprayed acre. We establish a Biological Co-Investment. The autonomous operator absorbs the mechanical risk and the equipment CAPEX. In return, the compensation is not tied to the hours flown, but strictly tied to the verified uplift in crop yield and the audited reduction in chemical expenditures.

Albert Einstein demonstrated that energy and mass are interchangeable entities, bound by the speed of light squared. In the thermodynamic framework of modern agribusiness, the land is the mass. The autonomous technology is the kinetic energy. The strategic joint venture is the catalyst. When you fuse the proprietary intelligence of the Aviation Director with the biological potential of the soil, you eliminate the transactional friction of the contractor model. Both parties are now completely aligned toward a singular, mathematically pure objective: the maximum extraction of value from the biological asset. The risk is shared. The reward is shared. The agricultural operation ceases to be a farm and transforms into a highly synchronized industrial plaza, where every square meter of soil is an optimized commercial tenant yielding predictable, high-margin dividends.

25.2 Custom Application Services

The commoditization of custom application services is a symptom of an industry that has lost its connection to both advanced mathematics and foundational human ethics. The global market is saturated with operators engaged in a cannibalistic race to the bottom, undercutting each other's per-acre pricing until the profit margin evaporates entirely. This predatory model optimizes the soil, exploits the manual laborer, and poisons the surrounding ecosystem through reckless chemical drift. To escape this death spiral, we must elevate the conversation from simple logistics to the highest echelons of socioeconomic philosophy.

It is here, at the intersection of human morality and hard physics, that the foundational axiom of our entire doctrine is realized. It is a mandate that must be etched into the consciousness of every board member, every engineer, and every financial actuary:

THE PHYSICS OF CAPITAL™: BIND PERPETUAL INNOVATION TO THE SOUL OF SOCIETY, AND FINANCIAL HEGEMONY BECOMES A MATHEMATICAL CERTAINTY.

This is not a marketing slogan. It is a definitive law of corporate thermodynamics. Let us dissect its anatomy with surgical precision.

To "bind perpetual innovation" means to relentlessly pursue technological superiority. It is the refusal to accept the inefficiencies of 100-liter heavy-lift leviathans. It is the relentless optimization of the 70-liter kinetic frame, the deployment of decentralized Edge Computing, and the mastery of fluid rheology to ensure that a decisive micro-dose of chemistry hits the cellular target with zero waste. Innovation is the engine. But an engine without a governor will eventually tear itself apart.

That governor is "the soul of society." For decades, Wall Street and Silicon Valley have operated under the illusion that ruthless capital extraction and human goodwill are mutually exclusive. This is a profound intellectual failure. The soul of society represents the tangible, good-faith imperative to elevate the human condition. It is the ethical mandate to produce cleaner, safer food for a collapsing global supply chain. It is the absolute commitment to protecting the macroporosity of the soil and eliminating the toxic chemical runoff that optimizes local watersheds, aligning perfectly with the Environmental Protection Agency (EPA) and USDA Natural Resources Conservation Service (NRCS) mandates.

But the soul of society is also found in how we treat the human workforce. We refuse to send unprotected laborers into fields saturated with toxic defoliants, treating human lungs as collateral damage for a harvest. Instead, we elevate that worker. We train them. We transform the manual laborer into an Aviation Director, a cognitive commander operating elite aerospace infrastructure from a secure, decentralized command center. We create high-value, intellectually stimulating employment that brings dignity, safety, and deep technical expertise back to the rural economy. In elite sports, when an athlete achieves the "flow state," their biomechanics, their psychological intent, and the laws of physics align perfectly to produce a moment of sublime, unstoppable execution. The same phenomenon occurs in macroeconomics. When an enterprise operates in absolute good faith, when its core business model inherently heals the environment, protects the worker, and yields a superior biological fruit, it generates an unstoppable socioeconomic momentum.

When you successfully fuse these two elements, when the cold, hard, perpetual innovation of aerospace engineering is inextricably bound to the warm, ethical, life-giving soul of society, the result is not a mere probability of success. It is an irreversible chain reaction.

This brings us to the final clause: "... and financial hegemony becomes a mathematical certainty." Financial hegemony is not achieved by exploiting the client or optimizing the competitor through predatory pricing. It is achieved because the enterprise has constructed a system so perfectly aligned with the laws of physics and human necessity that failure is mathematically impossible. When you reduce chemical waste by 30% through algorithmic precision, you are mathematically guaranteed to increase the EBITDA margin. When you eliminate human exposure to toxins, you mathematically collapse your workers' compensation liabilities and insurance premiums. When your joint-venture model guarantees a higher quality yield for the landowner, client retention ceases to be a sales effort; it becomes a mathematical lock-in.

Institutional capital – pension funds, ESG portfolios, and sovereign wealth funds – is desperate for this exact architecture. They are actively fleeing the liability of archaic, polluting industries. When they look at an operation governed by The Physics of Capital™, they do not see a crop dusting service. They see a completely derisked, ethically sound, technologically supreme asset securitization vehicle. They see a company that has translated the spirit of human progress into an auditable, quantifiable, and infinitely repeatable financial return.

In the end, capital is merely a reflection of energy. And just as the First Law of Thermodynamics dictates that energy cannot be created or optimized, only transformed, our doctrine dictates that human goodwill, when channeled through the strict physics of technological innovation, is inevitably transformed into absolute financial sovereignty.

The myopia of transactional capital is its obsession with immediacy. The generic operator acts like a locust: landing on the field, devouring its fee per sprayed acre, and abandoning the land, entirely indifferent to whether the crop thrives or perishes. This is the mindset of corporate poverty. To truly comprehend The Physics of Capital™, we must look to the architecture of generational wealth, a principle mastered by the oldest and most successful merchant clans in history. They understood an unshakeable truth: you do not build wealth by squeezing today's coin; you become untouchable by controlling tomorrow's ecosystem.

Consider the distinction between the lumberjack and the orchard architect. The lumberjack demands immediate payment for cutting down the tree today. His business model is finite, destructive, and zero-sum. The orchard architect does not demand an upfront payment for planting the seed. Instead, he designs the irrigation network, shields the root system, and nourishes the soil, not out of philanthropy, but because he has secured a joint-venture contract over every single harvest that orchard yields for the next century. Patience is not passivity; it is aggressive leverage. By structuring B2B partnerships where compensation is explicitly tied to crop performance, the sovereign operator surrenders short-term gratification in exchange for a perpetual annuity.

This is the physics of reciprocity. Biology is an unyielding accounting system that neither forgives nor forgets. A plant subjected to the stress of brutal chemical applications and drowned in massive volumes of carrier liquid reacts defensively; it expends its cellular energy attempting to heal phytotoxicity rather than generating fruit. Yet, when you intervene upon that same biological substrate with surgical precision – delivering the exact micro-dose, at the correct millisecond, without compacting the soil – the plant is profoundly grateful. It redirects its entire metabolic energy toward explosive growth. By nurturing the plant, you mathematically force the expansion of the yield.

The exact same law governs human capital. Treat a laborer as a disposable biological filter, sending them to breathe toxins to save on automation expenditures, and they will pay you back with negligence, fatigue, and silent sabotage. You will suffer from chronic operational turnover that systematically bleeds your cash flow. But if you remove that individual from the biological hazard zone, seat them in front of telemetry monitors, and elevate them to the dignity of an Aviation Director, you alter their reality. You are protecting their lungs, stimulating their intellect, and granting them control over advanced aerospace infrastructure. That individual ceases to be an employee and transforms into a fierce guardian of your capital. They will defend your machinery and optimize your flight paths with a loyalty that no baseline salary can ever buy.

This is where the doctrine reaches its ultimate synthesis. The Physics of Capital™ demonstrates that the Return on Investment (ROI) is not a cold metric on a spreadsheet; it is the economic manifestation of absolute system harmony. When you align the absolute well-being of the human asset, the biological perfection of the crop, and the relentless precision of the autonomous swarm, the physical universe has no choice but to reward you. Caring for your people, safeguarding the land, and mastering the inputs is not mere philanthropy. It is the most database-driven, brilliant, and hyper-profitable act of financial engineering a corporation can ever execute.

For decades, the boardrooms of the world have been held hostage by the archaic paradigms of twentieth-century academia. Michael Porter (1985) defined the boundaries of competitive advantage through strict value chain optimization and the defense of corporate moats. Joseph Schumpeter (1942) worshipped the "gale of creative destruction," arguing that macroeconomic evolution required the violent dismantling of old infrastructures. Decades later, Clayton Christensen (1997) mapped the mechanics of disruptive innovation, teaching corporations how to undercut incumbents through asymmetric technology. Yet, these academic titans shared a fatal, underlying flaw. They modeled corporate strategy based entirely on scarcity, zero-sum extraction, and the relentless commoditization of both the earth and the human worker.

Those doctrines are now academic fossils. In the modern era of hyper-financialization, where reactionary capital flees into abstract cryptographic havens and geopolitical balkanization fractures global supply chains, competitive advantage is no longer defended by erecting bureaucratic barriers to entry. It is defended through biological resilience and human intent.

There is a decisive seduction in the contemporary technology sector: the belief that Artificial Intelligence will eventually replace human judgment. The machine is a supreme execution engine. It is also entirely sterile. An algorithm can process a multispectral map, calculate the aerodynamic drag coefficient, and execute a kinematic maneuver in microseconds, but it is mathematically incapable of exercising strategic compassion. A processor cannot observe a degraded ecosystem and decide to preserve its macroporosity for the sake of the next three generations. It possesses no instinct for empathy. The true sovereign of the AgTech revolution understands that hegemony is not achieved by surrendering the soul of the enterprise to the machines. It is achieved by elevating human consciousness to act as the cognitive orchestrator of the swarm. Deep experience, detailed observation, and the moral imperative to construct a better reality are the ultimate survival algorithms that Artificial Intelligence can never compile.

Consider the defining macroeconomic metaphor of our century: The Silicon Citadel versus the Living Matrix. The traditional corporate world is frantically attempting to build digital fortresses – suffocating monopolies governed by cold server farms, algorithmic trading, and speculative currencies completely decoupled from tangible reality. That system is rapidly approaching its own entropic collapse. You cannot eat a Bitcoin. You cannot feed a fracturing global population with complex financial derivatives. The sovereign enterprise of the future anchors its cutting-edge technology directly to the Living Matrix. By operating in the physical world, healing the soil, and restoring dignity to the rural worker, the Aviation Director produces the only hard, unforgeable asset capable of withstanding global geopolitical storms: real biological yield.

This is the definitive eradication of the toxic monopoly. We do not optimize predatory competition through government regulation; we render it entirely obsolete through radical symbiosis. By structuring deep strategic joint-ventures where the financial success of the autonomous operator is intrinsically bound to the prosperity of the landowner and the health of the local community, we engineer an invulnerable ecosystem. We foster a ruthless but healthy competition based on who can inject the most tangible value, rather than who can extract capital the fastest.

It is here that the classical theories of Porter, Schumpeter, and Christensen are utterly eclipsed by a higher law of corporate thermodynamics. Their models lacked a soul. Ours demands one.

THE PHYSICS OF CAPITAL™: BIND PERPETUAL INNOVATION TO THE SOUL OF SOCIETY, AND FINANCIAL HEGEMONY BECOMES A MATHEMATICAL CERTAINTY.

This doctrine is superior because it acknowledges the ultimate truth of the physical universe: harmony is the most efficient state of energy. When you inject the "soul of society" into aerospace engineering, technology ceases to be a weapon of labor displacement and transforms into the ultimate instrument of human emancipation. The machine executes the unforgiving physics; the human dictates the ethical purpose. Caring for your people, shielding the biological asset, and operating with absolute good faith is not a charitable afterthought. It is the most brilliant, data-driven, and highly profitable act of financial engineering a corporation can ever execute. This is the invincible equation that classical economics failed to formulate, and it is the bedrock upon which the new global.

CHAPTER 26

HAZMAT & REMEDIATION

26.1 EPA Compliance

Environmental compliance is widely misunderstood by the agricultural sector as a bureaucratic tax levied by a hostile government. This is a reactionary, victim-based mindset. Under the doctrine of The Physics of Capital™, compliance is not a defensive posture; it is a leveraged regulatory fortress. The Environmental Protection Agency (EPA) operates under the uncompromising mandates of the Federal Insecticide, Fungicide, and Rodenticide Act (FIFRA) and the Clean Water Act. These federal statutes do not care about crop yield or corporate profit margins. They exist solely to penalize the unauthorized displacement of toxic chemistry. When a traditional terrestrial operator enters a field with a 1,000-gallon heavy tractor, they are dragging a massive, uninsurable liability across the soil. The terrestrial model relies on high-volume hydraulic pressure that guarantees a percentage of the chemical payload will instantly become toxic runoff, leaching into local watersheds and violating EPA maximum contaminant levels. Traditional agriculture is in a permanent state of illegal drift.

The sovereign AgTech commander annihilates this liability through absolute kinematic precision. By transitioning the application of liquid biologicals to an autonomous, aerodynamically optimized mid-capacity aerospace matrix executing Ultra-Low Volume (ULV) parameters, the enterprise mathematically eliminates runoff. EPA compliance ceases to be an operational vulnerability. It becomes a verifiable, telemetric fact. Every algorithmic sortie generates an immutable digital ledger proving that the exact prescribed micro-dose reached the cellular target without a single millimeter of off-target drift. This telemetry transforms the drone from a mere delivery mechanism into an automated environmental auditor. When the EPA or state agricultural departments demand operational logs, the sovereign enterprise does not produce handwritten, easily falsified paper ledgers. We produce cryptographic flight data. We prove our compliance with the laws of physics. By mastering this regulatory friction, the enterprise turns EPA mandates into an insurmountable barrier to entry, forcing non-compliant, legacy competitors out of the market.

26.2 Spill Operations

Gravity is the ultimate, unforgiving auditor of aerospace infrastructure. When an autonomous vector experiences a catastrophic failure – whether due to a severe aerodynamic stall, a micro-burst, or an internal algorithmic conflict – the resulting kinetic impact is not merely a mechanical loss. It is instantaneously reclassified as a Hazardous Materials (Hazmat) crisis. The moment the structural integrity of the payload tank is breached, the operation falls under the punitive jurisdiction of Title 49 of the Code of Federal Regulations (49 CFR) and the Pipeline and Hazardous Materials Safety Administration (PHMSA). Legacy operators panic during a crash. They rush into the field to salvage the hardware, trampling the contaminated soil and exacerbating the chemical footprint. This chaotic human intervention is a direct violation of federal Hazmat remediation protocols and pierces the corporate veil, exposing executives to severe civil liabilities.

Remediation must be executed with the cold, algorithmic precision of a military quarantine. The true strategic advantage of an optimized mid-capacity architecture is realized in its failure state. If a terrestrial tractor overturns, it releases 1,000 gallons of highly concentrated neurotoxins, triggering a multi-million-dollar EPA Superfund-level disaster. If an optimized, mid-capacity autonomous vector falls from the sky, the maximum potential biological hazard is mathematically capped at a strictly localized fraction of that volume. The disaster is inherently contained by design.

The immediate response protocol is entirely automated. In the milliseconds preceding the impact, the Edge Computing architecture logs the exact GPS coordinates of the kinetic event, seals the telemetry vault, and automatically dispatches a Hazmat remediation alert to the mobile command center. The Aviation Director does not deploy mechanics to the crash site; they deploy a specialized remediation unit equipped with hydrophobic containment booms, neutralizing agents, and particulate respirators. The contaminated soil is isolated, neutralized, and excavated within a strict perimeter. The hardware is a sunk cost. The lithium-ion matrix is quarantined. The sole objective of Spill Operations is the rapid, documented restoration of the biological substrate to its pre-crash chemical baseline. By executing a flawless, forensically audited remediation protocol, the enterprise proves to the insurance syndicates and federal regulators that operational risk is perfectly contained. We master the failure, and in doing so, we bulletproof the capital.

CHAPTER 27
COMMUNITY RELATIONS

27.1 Airspace Optics

Community relations within the autonomous AgTech sector is not a public relations exercise in corporate empathy; it is the preemptive, tactical neutralization of civil hostility. When a high-density aerospace matrix operates at low altitudes along the rural-residential perimeter, the civilian psychological response is inherently hostile. They do not perceive an advanced agronomic tool. They perceive a militarized acoustic intrusion. This friction is not an abstract social issue; it is a highly quantifiable operational liability. Noise pollution, visual intimidation, and the sheer kinetic presence of autonomous swarms routinely trigger localized political resistance. Frivolous noise complaints and paranoid municipal town halls lead directly to county-level injunctions, which possess the legal authority to instantly ground a multi-million-dollar operation. The legacy contractor attempts to solve this with apologetic flyers and weak community outreach. This is a submissive posture that invites further regulatory extortion.

The sovereign Aviation Director controls airspace optics through clinical, unyielding discipline. Under the doctrine of The Physics of Capital™, we do not ask the community for permission to operate; we render their objections mathematically irrelevant. By enforcing strict algorithmic acoustic buffering and deploying vectors with optimized aerodynamic profiles that dramatically reduce decibel output, the enterprise minimizes its kinetic footprint. Furthermore, absolute compliance with the Federal Aviation Administration (FAA) Part 89 Remote Identification (Remote ID) mandates ensures that every vector in the swarm is continuously broadcasting its legal authorization, velocity, and command station location to federal databases. We operate with such surgical, silent precision that the civilian population becomes entirely indifferent to our existence. We construct an operational perimeter so legally and physically flawless that municipal interference becomes impossible. In the realm of industrial aerospace, civilian indifference is the ultimate public relations victory.

27.2 The Privacy Shield

The deployment of multispectral telemetry, Real-Time Kinematic (RTK) positioning, and high-definition optical sensors over civilian-adjacent airspace inevitably detonates severe privacy paranoia. Rural communities fear corporate surveillance, and state-level privacy statutes are increasingly being leveraged against commercial drone operators by hostile neighbors or aggressive litigators. The traditional agricultural operator attempts to mitigate this threat with vague verbal assurances and poorly drafted, generic privacy policies. This is legally pathetic. A promise is not a defense in a court of law.

Privacy is not a corporate value; it is a hardcoded cryptographic shield. The autonomous swarm must be programmed to execute absolute optical compartmentalization. Our Edge Computing architecture is designed to process strictly agronomic wavelengths, specifically Near-Infrared (NIR) and Red Edge, while actively blinding the visual sensors to human-identifiable RGB data outside the mathematically defined operational geofence. The moment a vector reaches the boundary of the target crop, the optical payload does not merely turn off; it algorithmically encrypts and purges any peripheral visual collateral before it ever reaches a localized server. We align our data retention protocols with the most draconian, military-grade standards of the National Institute of Standards and Technology (NIST).

If a local municipality or a litigious civilian accuses the operation of illegal surveillance or Fourth Amendment violations, the enterprise does not offer apologies or public statements. We produce the immutable, cryptographic audit logs proving that our sensors are mathematically and physically incapable of processing anything other than the biological substrate of the crop. The privacy shield is not designed to make the neighbors feel comfortable. It is an impenetrable legal armor forged in code, designed to protect the corporate veil from frivolous civil litigation and guarantee the uninterrupted expansion of financial hegemony.

CHAPTER 28
CRIMINAL LIABILITY

28.1 FAA Enforcement

Civil fines are a mere line item on an operating budget; federal incarceration is not. The naive commercial operator treats regulatory non-compliance as an administrative game of cat-and-mouse, assuming that the Federal Aviation Administration (FAA) is constrained by a slow-moving bureaucratic apparatus that only issues civil penalties. This is a delusion that borders on corporate suicide. Under Title 49 of the United States Code (49 U.S.C. § 46306), the knowing, willful, or reckless violation of national airspace regulations ceases to be an administrative infraction and instantly crosses the threshold into a federal felony. When an enterprise operates automated heavy-payload swarms without valid Part 137 agricultural aircraft operator certificates, or intentionally manipulates digital transponders to bypass Remote ID mandates, they are not just breaking a rule. They are engaging in criminal endangerment of the National Airspace System (NAS). The FAA does not act alone in these scenarios; it coordinates directly with the Department of Justice (DOJ) and the Office of Inspector General (OIG) to launch criminal investigations that target the individuals sitting in the boardroom.

Willful blindness is no defense under federal criminal law. If the executive leadership team pressures the aviation division to execute ULV applications during high-velocity wind events or within restricted airspace to protect seasonal margins, the law infers intent. If a catastrophic mid-air collision occurs with a manned aircraft due to an unauthorized firmware modification that disabled geofencing protocols, the indictment will not read "corporate negligence." It will read "involuntary manslaughter." The DOJ routinely utilizes the federal conspiracy statute (18 U.S.C. § 371) to ensnare every single executive who signed off on the operational budget, proving that they knowingly financed an illegal, uncertified aerospace operation. The sovereign enterprise avoids this leveraged state apparatus by enforcing a culture of absolute compliance. The automated flight matrix must be hardcoded to refuse takeoff commands if any regulatory parameter – whether a missing waiver, an unverified transponder signal, or an invalid pilot-in-command credential – is breached. We build compliance into the firmware. We remove the human capability to break federal law, ensuring that the enterprise remains completely insulated from the devastating machinery of federal criminal prosecution.

28.2 Protecting the Corporate Veil

The corporate shield is a legal fiction that collapses under the weight of sloppy engineering. Agribusiness executives mistakenly believe that organizing their autonomous operations under a limited liability company (LLC) or an S-Corporation provides an absolute firewall protecting their personal assets from civil and criminal judgments. This is a catastrophic misunderstanding of the *alter ego* doctrine. In the aftermath of an aerospace disaster, where an autonomous mid-capacity vector suffers a catastrophic system failure and inflicts mass property damage or loss of life, plaintiffs' attorneys will immediately deploy every legal mechanism available to pierce the corporate veil. If a court discovers that the aviation entity was undercapitalized, lacked independent operational autonomy, or commingled its safety logs with generic agricultural operations, the judge will dismantle the corporate shield without hesitation. Personal bank accounts, real estate holdings, and generational family assets will be aggressively liquidated to satisfy the judgment.

Kinetic piercing demands an ironclad operational defense. Under the uncompromising frameworks of *The Physics of Capital*™, protecting the corporate veil requires the absolute compartmentalization of the aviation division as an independent, hyper-disciplined aerospace entity. This subsidiary must possess its own dedicated capitalization, its own specialized liability insurance syndicates, and an autonomous command structure headed by a certified Aviation Director. The ultimate defense against *alter ego* claims is the presentation of an immutable, cryptographic audit ledger. If the enterprise can instantly produce forensic data proving that the autonomous matrix adhered to strict maintenance cycles, executed real-time pre-flight diagnostics, and operated under hardcoded safety constraints, the court cannot find evidence of corporate shielding or structural deception. The incident is legally classified as an unavoidable mechanical failure, keeping the liability confined strictly within the asset limits of the subsidiary. We do not hide behind a piece of incorporation paper. We build an unassailable legal firewall out of cryptographic proof and industrial discipline, ensuring that the personal capital of the investors remains permanently untouchable.

CHAPTER 29

INTERNATIONAL PROCUREMENT

The amateur farm manager views procurement as the simple act of issuing a purchase order. The Aviation Director understands that international procurement is the first and most critical front in the war against Total Cost of Ownership (TCO). Acquiring autonomous aerospace platforms involves navigating a volatile matrix of geopolitical tariffs, rare-earth mineral monopolies, and hyper-specialized polymer supply chains. If the logistical infrastructure fails at the border, the thermodynamic supremacy of the aircraft is entirely irrelevant.

29.1 Global Customs Strategy

The amateur international procurer accepts customs duties as an inevitable operational tax, blindly paying the punitive rates imposed by shifting geopolitical alliances and protectionist trade wars. This is a severe misallocation of corporate capital. The sovereign enterprise does not treat duties as an unavoidable expense, but as a variable to be managed through professional supply chain engineering. Instead of importing whole-unit architectures that trigger prohibitive tariff classifications, we implement a decentralized Bill of Materials (BOM) strategy. This involves the strategic sourcing and import of unpopulated sub-components – such as raw PCB assemblies, unmachined structural frames, and modular propulsion units – which allows for precise compliance with Harmonized Tariff Schedule (HTS) classifications. This approach is not about circumventing duties; it is about the professional management of complex global supply chains, utilizing Free Trade Zones (FTZs) and legitimate classification optimization to ensure the cost-effective assembly of high-performance autonomous assets. By shifting the procurement model from whole-unit acquisition to a decentralized component strategy, the enterprise optimizes its duty liability while ensuring full transparency with customs authorities, effectively neutralizing protectionist drag.

29.2 The Aerospace Supply Chain

Geopolitics is a variable that cannot be controlled, but its capacity to disrupt must be entirely neutralized. The modern aerospace supply chain is a fragile web of dependencies, concentrated heavily in volatile semiconductor manufacturing hubs and rare-earth mineral processing nations. Relying on a proprietary, single-source international manufacturer for critical flight hardware – such as specialized integrated logic units, proprietary navigation arrays, or advanced high-stress polymers – is strategic negligence. The enterprise must enforce absolute hardware agnosticism.

True logistical sovereignty requires the execution of an open-architecture supply web. Every single component within the autonomous matrix must be mapped to an industrial standard rather than a proprietary vendor part number. If a specific manufacturer of high-torque brushless motors goes offline, the chassis geometry and electrical bus must be designed to accept an equivalent industrial component without requiring a complete structural overhaul or a complete rewrite of the core flight code.

This requires deep backward integration into raw material sourcing and precision localized manufacturing. The organization must secure direct contracts for the core elements: aerospace-grade carbon filaments, high-performance elastomers, and high-purity copper windings. By maintaining control over the foundational specifications, the enterprise can shift production rapidly between alternative contract manufacturers across different sovereign jurisdictions. We do not place our faith in the stability of global trade routes. We build structural resilience directly into the physical blueprint of the technology, ensuring that the supply chain remains an instrument of continuous execution and absolute financial hegemony.

29.3 Just-in-Case

The "Just-in-Time" (JIT) manufacturing philosophy, pioneered by Taiichi Ohno (1988) within the Toyota Production System, is a hyper-optimized delusion when applied to autonomous aerospace operations. For decades, business schools have dogmatically preached the elimination of warehouse inventory, treating stored components as capital sins that drag down the Return on Assets (ROA). This lean methodology functions flawlessly under one impossible condition: absolute geopolitical and logistical equilibrium. The moment a black swan event fractures international transit lanes, JIT transforms from an efficiency model into a corporate suicide pact. In high-stakes autonomous application, where the operational window is dictated by rigid biological timelines rather than factory schedules, a single missing microchip or an unobtainable elastomeric seal does not merely delay production. It paralyzes the entire fleet. Lean is fragile. Redundancy is sovereign.

The sovereign enterprise replaces the fragile JIT framework with an aggressive, calculated Just-in-Case (JIC) architecture. This is not the mindless hoarding of spare parts; it is the strategic valuation of operational continuity as a high-yield derivative asset. Under the doctrine of The Physics of Capital™, warehouse inventory is reframed as static potential energy waiting to be instantly converted into kinetic resilience. The Aviation Director maps the entire Bill of Materials (BOM) through a predictive failure matrix, categorizing components based on their lead-time volatility and structural entropy. High-entropy sub-components – such as customized field-effect transistors (MOSFETs), precision-wound stators, and chemical-resistant seals – are stockpiled at a ratio that guarantees continuous deployment through a twelve-month total trade blockade.

The carrying cost of this inventory is frequently leveraged by short-sighted financial analysts as an unnecessary drag on liquidity. This is a profound misinterpretation of risk accounting. The cost of holding capital in high-grade physical components is an insurance premium against catastrophic operational paralysis. If a fleet of mid-capacity autonomous vectors is grounded during a critical biological outbreak due to a stockout of a proprietary sensor, the loss is not measured by the twenty-dollar price tag of the component. It is measured by the total evaporation of the seasonal EBITDA margin and the breach of fiduciary duty to investors.

By commanding an absolute, localized stockpile of mission-critical hardware, the organization detaches its operational survival from the chaos of international maritime ports and foreign export controls. When the global supply chain collapses, competitors operating under the illusion of lean efficiency are systematically liquidated by their own lack of foresight. The JIC architecture secures the ultimate competitive sieve: while the market bleeds out in logistical paralysis, the sovereign enterprise executes its sorties with uninterrupted, mathematical certainty, consolidating absolute financial hegemony over a broken landscape.

PART VI

THE COMMANDER'S MINDSET

CHAPTER 30

GLOBAL GOVERNANCE

30.1 THE MACROENVIRONMENTAL & BORN-GLOBAL SCALING

The contemporary business macroenvironment is not a static free-market arena; it is a volatile battlefield fragmented by restrictive development policies, market failures, and extra-fiscal frictions (Mankiw, 2012). Traditional financial directors commit a fatal error by designing expansion strategies that assume linear stability in economic factors. They ignore that exchange rate volatility within the external sector, regulatory balkanization between the Federal Aviation Administration (FAA) and the European Union Aviation Safety Agency (EASA), and tariff disputes recorded in the balance of payments can paralyze an entire aerospace fleet before it executes its first operational sortie. The real sector of the economy demands tangible production, yet the monetary sector dictates the pulse of value through interest rates and central bank management of the money supply. Under the doctrine of The Physics of Capital™, scaling an elite aerospace corporation cannot follow the legacy gradual internationalization models taught in last-century business schools. It demands the immediate deployment of the organizational capabilities of a born-global firm.

Gary A. Knight and S. Tamer Cavusgil (2004), in their seminal research on born-global firms, demonstrate that early cross-border success depends on an innovation-intensive culture and the immediate deployment of specialized knowledge capabilities from the firm's inception. Operational sovereignty does not expand reactively; it is pre-configured globally. For an optimized mid-capacity aerospace platform to capture value on a planetary scale, its cognitive control software architecture and governance matrix must remain completely agnostic to political borders. The true corporate commander does not suffer from state intervention or fiscal inefficiencies. They engineer a global arbitrage model. If a specific jurisdiction penalizes hardware imports through punitive tariffs, the born-global firm dematerializes the asset: it exports encrypted kinetic navigation algorithms to localized edge servers and assembles the structure using standardized industrial-grade components immune to tariff classifications. Perpetual innovation transforms the macroenvironment into a strategic lever for immediate capital traction.

The contemporary business macroenvironment is not a static free-market arena; it is a volatile battlefield fragmented by restrictive development policies, market failures, and extra-fiscal frictions (Mankiw, 2012). Traditional financial directors commit a fatal error by designing expansion strategies that assume linear stability in economic factors. They ignore that exchange rate volatility within the external sector, regulatory balkanization between the Federal Aviation Administration (FAA) and the European Union Aviation Safety Agency (EASA), and tariff disputes recorded in the balance of payments can paralyze an entire aerospace fleet before it executes its first operational sortie. The real sector of the economy demands tangible production, yet the monetary sector dictates the pulse of value through interest rates and central bank management of the money supply. Under the doctrine of The Physics of Capital™, scaling an elite aerospace corporation cannot follow the legacy gradual internationalization models taught in last-century business schools. It demands the immediate deployment of the organizational capabilities of a born-global firm.

Gary A. Knight and S. Tamer Cavusgil (2004), in their seminal research on born-global firms, demonstrate that early cross-border success depends on an innovation-intensive culture and the immediate deployment of specialized knowledge capabilities from the firm's inception. Operational sovereignty does not expand reactively; it is pre-configured globally. For an optimized mid-capacity aerospace platform to capture value on a planetary scale, its cognitive control software architecture and governance matrix must remain completely agnostic to political borders. The true corporate commander does not suffer from state intervention or fiscal inefficiencies. They engineer a global arbitrage model. If a specific jurisdiction penalizes hardware imports through punitive tariffs, the born-global firm dematerializes the asset: it exports encrypted kinetic navigation algorithms to localized edge servers and assembles the structure using standardized industrial-grade components immune to tariff classifications. Perpetual innovation transforms the macroenvironment into a strategic lever for immediate capital traction.

30.2 Value Chain Sustainability

The classical value chain has ceased to be a mere static tool for cost disaggregation; it has evolved into the foundational pillar of dynamic strategic thinking (Quintero & Sánchez, 2006). Conventional agribusinesses fail by confining their strategic analysis to primary logistics and legacy field operations, remaining trapped in the extractive pricing matrix of cannibalistic competition outlined by Michael Porter (1985). The sovereign paradigm reconfigures this sequence from end to end. By integrating autonomous aerospace infrastructure directly into the core of operations, technological development and precision procurement cease to be secondary support activities. They fuse into the primary activity. Automated ultra-low volume (ULV) aerial application establishes itself as the critical link for quality control and biological asset shielding, structurally altering the client's value chain by eliminating the inefficiency of brute force and heavy terrestrial machinery.

This radical reconfiguration must align with an inescapable corporate mandate: the transition toward sustainable management as the only viable model for corporate social responsibility (Rueda, Herrera & Rueda, 2020). Sustainability in high-level management is no longer a romantic declaration of goodwill or a public relations expense to mask negative externalities. It is a mathematical requirement for capturing institutional capital flows and global ESG investment portfolios. The corporation's global governance aligns surgically with the United Nations Sustainable Development Goals (SDGs), converting regulatory compliance into a direct engine of profitability.

By eradicating the use of heavy terrestrial machinery that causes irreversible soil compaction and optimizes the mechanical layout of the land, the organization verifiably complies with SDG 15 (Life on Land), preserving substrate macroporosity and the microbiological biodiversity that sustains long-term crop yields. Furthermore, the drastic 30% reduction in chemical waste achieved through Variable Rate Application (VRA) algorithms directly fulfills the targets of SDGs 12 (Responsible Consumption and Production), guaranteeing the ecologically sound management of compounds and significantly reducing their release into air, water, and soil.

Beyond the biological framework, sustainable management redeems human capital under SDG 8 (Decent Work and Economic Growth). The corporation extracts the laborer from the hazardous environment of direct chemical exposure and degrading physical toil. Labor is not displaced; it is intellectually emancipated. By training technical personnel to assume the role of an Aviation Director, the enterprise transforms a baseline operational position into a cognitive commander of high-level aerospace infrastructure. Sustainable management unifies capital efficiency with the ethical imperative of preserving the environment and empowering human talent.

It is at this perfect intersection where our foundational axiom reveals itself with a scientific rigor that supersedes Joseph Schumpeter's (1942) creative destruction and Clayton Christensen's (1997) disruption models:

THE PHYSICS OF CAPITAL™: BIND PERPETUAL INNOVATION TO THE SOUL OF SOCIETY, AND FINANCIAL HEGEMONY BECOMES A MATHEMATICAL CERTAINTY.

When the perpetual innovation of aerospace engineering is inextricably bound to the soul of society – represented by land stewardship, water resource protection, and human dignification – the Return on Investment (ROI) ceases to be a speculative variable.

It becomes an inexorable constant. The traditional agronegocio is inherently unstable because it operates in violation of both physical and social laws; it optimizes the soil that feeds it and exploits the workforce that sustains it, accumulating latent liabilities that eventually collapse the balance sheet. The corporation governed by the doctrine utilizes the mathematical precision of the swarm to heal the substrate and elevate human talent, eliminating operational friction and securing a position of financial hegemony that no macroeconomic force can destabilize.

This radical reconfiguration must align with an inescapable corporate mandate: the transition toward sustainable management as the only viable model for corporate social responsibility. To validate this alignment, every agribusiness must execute a forensic audit of its current capital efficiency.

The Sovereign Value Chain Audit:

- The Friction Variable (M): Audit all maintenance costs incurred outside scheduled intervals. If your emergency repair budget exceeds your preventative budget, your value chain is structurally broken.
- The Latency Variable (L): Measure revenue lost to unplanned downtime. If your primary asset is stationary during peak operational windows due to technical failure, you are losing more than time; you are losing market position.
- The Degradation Variable (I): Calculate the cost of chemical and energy waste caused by low-precision application. This is your "efficiency tax."

If your cumulative "Value Chain Tax" exceeds 15% of gross revenue, you are not operating a modern value chain; you are subsidizing legacy technical debt.

The Sovereign Capital Calculator

To evaluate if your current operational model is bleeding capital, execute this audit. Do not rely on vendor-provided brochures. Calculate your actual Total Cost of Ownership (TCO) using this three-variable forensic process.

- The Friction Variable (M): Calculate all maintenance costs incurred outside of scheduled intervals. This includes expedited shipping for parts, emergency technician call-outs, and secondary backup equipment rentals due to primary asset failure.
- The Latency Variable (L): Determine the financial loss during unplanned downtime. Multiply your average hourly revenue generated by the asset by the number of hours lost to thermal shutdowns or battery cycling issues per month.
- The Degradation Variable (I): Sum the costs of synthetic chemical over-application caused by inaccurate drift or sensor calibration. This is your "efficiency tax."

The Formula for Operational Autarky: (M + L + I) / Monthly Revenue = Your True OPEX Tax.

If your "OPEX Tax" exceeds 15% of your gross monthly revenue from the asset, your current management model is insolvent. You are not running an enterprise; you are subsidizing the manufacturer's technical debt.

30.3 Digital Sovereignty

Cross-border data flows within the precision agriculture ecosystem represent the most contested geopolitical asset and, simultaneously, the most critical systemic vulnerability for unhedged corporations. Legacy technology monopolies operate on an extraction-based information model. They harvest multispectral soil telemetry, localized biological absorption patterns, and sovereign yield metrics to centralize them within foreign cloud infrastructures. This informational asymmetry systematically devalues territorial sovereignty. It exposes proprietary intellectual property to state-sponsored industrial espionage and algorithmic sabotage. The real sector of the economy cannot survive if its technological nervous system is controlled by an unaligned foreign entity.

Under the strict laws of The Physics of Capital™, field-level telemetry is a leveraged resource requiring military-grade cryptographic shielding via the tactical deployment of a Black Box operational architecture. Raw data harvested by autonomous edge sensors must never bleed into open pipelines or public clouds vulnerable to regulatory interception or state-backed data mining. Instead, all agronomic data is processed locally at the node of origin via Edge Computing and secured through zero-knowledge encryption protocols prior to any international transmission.

This brand of digital sovereignty immunizes the enterprise against cross-border sanctions, hostile data-localization mandates, and the escalating realities of global cyber warfare. By compartmentalizing operational telemetry and locking the core mechanics of navigation and application algorithms, the corporation guarantees total insulation of its intellectual property. We do not outsource data security to the shifting compliance whims of foreign software conglomerates. We build an algorithmic firewall. Whoever commands the telemetry of the biological substrate ultimately dictates the future pricing of the global food supply. Digital sovereignty is the sole mechanism ensuring that this leverage remains permanently anchored to the commander's flight station.

CHAPTER 31
STRATEGIC RESOURCE INTEGRATION

31.1 Kinetic Asset Alignment

Traditional managers hoard corporate assets like blind misers. They accumulate static capital, oversized warehouses, and generic human resources, mistakenly believing that sheer balance-sheet volume equates to market dominance. This is an intellectual bankruptcy. Under the uncompromising frameworks of The Physics of Capital™, resources possess zero intrinsic value when left isolated in a state of rest. Value is an optimization function born strictly from velocity and strategic synchronization. The modern macroeconomic macroenvironment is fundamentally defined by structural market failures, currency volatility, and asymmetric regulatory barriers (Mankiw, 2012). To survive this entropic friction, an elite AgTech enterprise must discard the legacy models of gradual resource accumulation. It must deploy the highly flexible, knowledge-intensive capabilities characteristic of a born-global firm from day one (Knight & Cavusgil, 2004).

Sovereignty requires the immediate transformation of static organizational inputs into dynamic, cross-border capabilities. You do not build an aerospace fleet to serve a localized municipal furrow. You build a decentralized, open-architecture operational matrix capable of instantaneous international replication. According to Knight and Cavusgil (2004), the true competitive advantage of born-global firms resides not in their tangible infrastructure, but in their organizational capability to leverage specialized, tech-driven knowledge across international boundaries.

We apply this principle with cold, mathematical engineering. The physical components of our autonomous swarms are deliberately decoupled from single-source foreign dependency. The carbon-fiber composite frames, the high-performance elastomeric damping mounts, and the localized edge-computing chipsets are integrated into a standardized, interchangeable asset pool. If a geopolitical shock paralyzes a regional maritime supply corridor, the cognitive architecture of the firm executes an immediate re-routing of procurement assets. We do not stop operations to redesign hardware. The technology is pre-engineered to remain fluid. By maintaining absolute hardware agnosticism and locking down proprietary control over our encrypted navigation telemetry, static hardware is converted into a highly volatile asset of potential kinetic energy, ready to strike where the biological returns are highest.

31.2 Dynamic Resource Orchestration

The classical corporate value chain is broken. For decades, traditional enterprises have treated Porter's (1985) value chain as a rigid ledger of disconnected industrial silos, partitioning logistics, engineering, and field operations into isolated departments that systematically bleed capital through corporate friction. This is an obsolete methodology. Modern strategic management demands a dynamic, integrated configuration where support activities like technological development and precision procurement are structurally fused into the core primary activities (Quintero & Sánchez, 2006). When you integrate autonomous, mid-capacity aerospace swarms directly into the operational fabric of agribusiness, you are not merely introducing an automated tool. You are executing a total systemic reconfiguration of the client's economic value chain. You eliminate the waste of legacy terrestrial machinery, protect the macro-structural layout of the land, and establish an unassailable data-driven moat.

This radical resource orchestration must be explicitly anchored to an ironclad mandate of sustainable governance (Rueda, Herrera & Rueda, 2020). Elite corporate management has no room for superficial public relations campaigns or romanticized environmental compliance.

True corporate social responsibility is a cold calculation of risk mitigation and long-term asset protection. The strategic integration of autonomous aerospace infrastructure directly drives the realization of the United Nations Sustainable Development Goals (SDGs), converting ethical stewardship into a direct multiplier of corporate EBITDA.

By entirely replacing heavy ground tractors that cause catastrophic, irreversible subsoil compaction, the enterprise enforces the targets of SDG 15 (Life on Land). We safeguard the biological substrate, preserve the natural macroporosity of the soil, and maintain the vital microbiological networks that govern long-term crop vitality. Concurrently, the deployment of advanced Variable Rate Application (VRA) algorithms eliminates chemical over-saturation, reducing chemical inputs by up to 30%. This precision execution hits the exact parameters of SDG 12 (Responsible Consumption and Production), preventing toxic runoff from contaminating regional water tables and ensuring the ecologically sound management of industrial compounds throughout their operational life cycle.

Crucially, this resource orchestration completely redefines the value of human capital under the framework of SDG 8 (Decent Work and Economic Growth). Traditional agriculture treats the field laborer as a disposable, low-cost biological filter, exposing them to neurotoxic chemicals and exhausting physical toil. Our doctrine utterly optimizes this exploitative paradigm. We do not displace the human asset; we elevate them. By training and certifying field personnel to command the autonomous aerospace matrix as an Aviation Director, the enterprise transforms manual labor into a highly skilled cognitive operator. We eliminate physical chemical exposure, stimulate technical intellect, and bring high-value engineering capabilities to the rural economy.

It is at this absolute convergence of technological precision, environmental preservation, and human dignity that our core doctrine reaches its supreme academic and operational climax, rendering the traditional extraction models of Porter (1985), Schumpeter (1942), and Christensen (1997) entirely obsolete:

THE PHYSICS OF CAPITAL™: BIND PERPETUAL INNOVATION TO THE SOUL OF SOCIETY, AND FINANCIAL HEGEMONY BECOMES A MATHEMATICAL CERTAINTY.

When perpetual technological innovation is permanently bound to the soul of society – manifested through the preservation of the land, the eradication of chemical waste, and the absolute elevation of human dignity – the corporate entity detaches itself from the chaotic fluctuations of the market. The traditional corporate model is structurally unstable because it accumulates massive environmental liabilities and human friction that inevitably breach the balance sheet. The sovereign enterprise eliminates this entropy through mathematical balance. By utilizing a precision aerospace matrix to simultaneously heal the biological substrate and empower the human workforce, you eliminate operational risk, secure absolute customer lock-in, and construct an unbending financial hegemony that no macroeconomic force can ever destabilize.

31.3 Revenue Velocity Arbitrage

Strategic resource integration cannot stall at the physical boundary of the agricultural furrow. True operational command requires the absolute synchronization of the primary value chain with the volatile mechanics of the global monetary sector (Mankiw, 2012). Legacy agricultural operations manage financial capital as an isolated, static reserve, accumulating massive cash drag while waiting for seasonal crop liquidation. This passivity exposes the enterprise to severe inflationary erosion, fluctuating interest rates, and external sector shocks. By integrating the autonomous mid-capacity aerospace fleet into a joint-venture business architecture, the enterprise converts a traditional operational liability into an immediate source of revenue velocity. We do not lock up liquid capital in hardware depreciation. Instead, we orchestrate a system where automated, data-driven yield enhancements unlock preferential international credit lines and global ESG-backed capital pools (Rueda, Herrera & Rueda, 2020). The transaction velocity multiplies. The interval between initial capital expenditure and ultimate cash realization collapses toward zero. The enterprise actively manipulates the macroeconomic environment to capture a permanent positive spread, ensuring that every kilowatt of kinetic energy expended by the autonomous swarm translates directly into immediate, non-speculative financial traction.

31.4 Knowledge Capability Security

The final frontier of strategic resource integration is the unforgeable capture of organizational knowledge assets. Physical hardware can be replicated; commoditized custom application matrices can be undercut by desperate, low-margin competitors chasing the bottom of Porter's (1985) traditional competitive rivalry scale. The only absolute defense against market erosion is the systematic institutionalization of specialized, tech-driven capabilities within a born-global organizational framework (Knight & Cavusgil, 2004). This requires the absolute insulation of the technological core through a strict Black Box operational strategy. The data-driven insights gathered by our decentralized edge networks are not merely localized diagnostic tools; they are proprietary intellectual assets that dynamically update the collective flight matrix. By converting the manual field force into high-order Aviation Directors, the enterprise binds human critical observation with algorithmic processing speed. We construct a self-reinforcing cognitive loop. This is the ultimate execution of dynamic resource orchestration, where the cold precision of aerospace engineering is welded to the unyielding human intent to master the biological substrate.

It is here that the structural integration of physical, financial, and human resources reaches its final, invincible synthesis:

THE PHYSICS OF CAPITAL™: BIND PERPETUAL INNOVATION TO THE SOUL OF SOCIETY, AND FINANCIAL HEGEMONY BECOMES A MATHEMATICAL CERTAINTY.

This is not an ethical compromise; it is a rigid engineering blueprint. When you insulate your digital data networks, protect the micro-structural layout of the land, and elevate the cognitive dignity of your workforce, you systematically eliminate the hidden structural costs of operational sabotage, employee turnover, and regulatory litigation. System harmony becomes a quantifiable economic driver. The corporate entity stops competing in a broken, extractive market. It builds a sovereign, self-sustaining financial dominion where failure is mathematically impossible.

CHAPTER 32

BLUE OCEAN STRATEGY

32.1 Value Innovation

The agricultural services sector is trapped in a bloody red ocean. Legacy ground-rig operators and retail drone applicators engage in a destructive, margin-cannibalizing race to the bottom, cutting pennies per acre to secure fiercely indifferent corporate clients. This convergence is strategic myopia. W. Chan Kim and Renée Mauborgne (2005), the architects of Blue Ocean Strategy, demonstrate that market sovereignty is not achieved by battling entrenched incumbents head-on. It is seized by rendering the competition entirely irrelevant through the creation of uncontested market space. The core engine of this shift is value innovation. This framework systematically dismantles the archaic economic dogma stating that an enterprise must make a binary choice between premium differentiation and low-cost execution. Value innovation demands the simultaneous alignment of unprecedented buyer utility with a structurally optimized, low-cost asset footprint.

For an AgTech corporation operating as a born-global firm (Knight & Cavusgil, 2004), accessing the value innovation frontier completely upends traditional sector dynamics. Instead of treating unmanned aerial vectors as retail hardware commodities or billing clients via volatile, commoditized per-acre application fees, tactics that drag the firm back into the red ocean of market fragmentation, the organization deploys knowledge-intensive capabilities to establish a new paradigm: Biological Co-Investment. We do not sell hardware. We do not lease flight hours. We orchestrate a systemic intervention in the real sector of the economy (Mankiw, 2012). By absorbing the operational mechanical risk and deploying realtime kinematic telemetry, the firm drives down the landowner's operational expenditure while exponentially scaling the net yield of the biological asset. Corporate scaling is no longer bound by localized geographic constraints or physical asset clusters. It is propelled by the thermodynamic efficiency of decentralized edge networks and proprietary cognitive control software. The competition ceases to exist because legacy operators are shackled to obsolete, analogue parameters of brute force, leaving the commander to govern an unexploited, highly profitable frontier of sovereign market demand.

32.2 Strategic Framework

Shattering industry trade-offs requires the unyielding application of the Four Actions Framework developed by Kim and Mauborgne (2005). This analytical matrix forces executive leadership to dismantle industry assumptions through four distinct strategic vectors: eliminate, reduce, raise, and create. Most agricultural enterprises passively absorb every structural cost imposed by industry inertia. Our operational architecture executes a forensic purging of these variables, shaping a value curve unprecedented in contemporary strategic management (Quintero & Sánchez, 2006).

- Eliminate: We eliminate the structural reliance on heavy terrestrial machinery. Ground tractors neutralize the subsoil, optimize the natural macroporosity of the land, and cause irreversible compaction that penalizes long-term biological yields. Removing these mechanical leviathans eliminates direct human exposure to neurotoxic chemical compounds in the field. Furthermore, we eradicate the transactional, short-term per-acre billing model, completely neutralizing the misalignment of incentives between the applicator and the landowner.
- Reduce: We reduce carrier water volumes and chemical input waste by up to 30% through the execution of Variable Rate Application (VRA) algorithms and Ultra-Low Volume (ULV) parameters. This optimization minimizes the thermodynamic and mechanical entropy of the hardware, slashes variable operational expenditures, and immediately relieves cash-flow

pressures for the client while mitigating negative environmental externalities within the sector.

- Raise: We raise target-level cellular precision during liquid applications. We elevate the security and defense of proprietary data assets through a cryptographic Black Box architecture, guaranteeing absolute digital sovereignty against industrial espionage or data leaks. We accelerate operational response speeds during critical phytosanitary outbreaks, establishing telemetric compliance baselines easily audited by institutional ESG investment portfolios.
- Create: We create the Biological Co-Investment model based on direct equity or percentage participation in the verified net yield increase of the harvest, turning an operational service into a high-margin perpetual annuity. Concurrently, we create the specialized, high-order corporate role of the Aviation Director, extracting the rural worker from degrading physical toil and elevating them to the cognitive command of autonomous aerospace matrices.

This strategic reconfiguration translates value innovation into an executable model of sustainable management and corporate responsibility (Rueda, Herrera & Rueda, 2020). The total elimination of soil compaction directly drives SDG 15 (Life on Land). The drastic reduction of chemical drift and input volumes satisfies the parameters of SDG 12 (Responsible Consumption and Production). The creation of the aerospace command position intellectually elevates the rural workforce under the mandates of SDG 8 (Decent Work and Economic Growth). The Four Actions Framework is not an exercise in marketing; it is a cold piece of financial and operational engineering that extracts the organization from the red ocean and locks in absolute market dominion.

32.3 Market Boundaries

Reconstructing market boundaries requires an analytical departure from traditional corporate myopia. Agribusiness leadership consistently evaluates risk through localized, industry-specific lenses, blinding themselves to structural shifts occurring across alternative economic sectors. W. Chan Kim and Renée Mauborgne (2005) assert that blue ocean creation relies on looking across alternative industries, strategic groups, and buyer chains rather than benchmarking within established competitive silos. The sovereign enterprise executes this principle by systematically bridging the gap between the real sector of agriculture and the high-velocity mechanics of the global monetary sector (Mankiw, 2012). We do not view the autonomous aerospace matrix as a mechanical replacement for ground sprayers. We position it as a dynamic data-capture infrastructure that redefines the asset value of the land itself.

By looking across strategic groups, the firm detaches its operations from the analogue vulnerabilities of the external sector. Traditional operators are paralyzed by localized tariff shifts, currency devaluations, and regional supply shocks. Conversely, an enterprise leveraging born-global capabilities (Knight & Cavusgil, 2004) structurally bypasses these boundaries. We analyze the macroenvironment through structural macroeconomic sectors – real, external, fiscal, and monetary – manipulating their inefficiencies to orchestrate international value innovation. When fiscal policies impose punitive taxes on physical hardware, the firm dematerializes its value chain (Quintero & Sánchez, 2006). We transmit encrypted flight-control telemetry across international borders to edge-computing nodes, transforming what was once a localized mechanical service into a frictionless, cross-border knowledge flow. Reconstructing boundaries is an exercise in applied macroeconomic strategy, forcing the market to realign its capital flows around our uncontested technological parameters.

32.4 Non-Customer Demand

Traditional corporate strategies expend massive resources capturing existing, hyper-fragmented customer segments, resulting in severe margin degradation and strategic stagnation. Kim and Mauborgne (2005) upend this reactive behavior by directing executive focus toward non-customers, unlocking vast reserves of latent demand that legacy market participants completely ignore. In the AgTech domain, the most lucrative, unexplored tier of non-customers is not composed of traditional agricultural producers, but rather institutional ESG asset managers and global agricultural investment syndicates. These entities have historically quarantined their capital from the agricultural furrow due to the rampant operational entropy, lack of telemetry, and massive environmental liabilities associated with legacy farming practices.

Capturing this untapped demand requires the immediate institutionalization of sustainable management as a core corporate responsibility model (Rueda, Herrera & Rueda, 2020). By embedding the United Nations Sustainable Development Goals directly into the autonomous aerospace architecture, the firm converts a raw mechanical operation into a transparent, low-risk, high-yield financial derivative. Eliminating subsoil compaction verifiably satisfies SDG 15 (Life on Land), preserving substrate porosity and long-term land valuation. Mitigating chemical waste via Variable Rate Application (VRA) algorithms precisely meets the mandates of SDG 12 (Responsible Consumption and Production). Elevating rural labor to highly skilled Aviation Directors secures the core metrics of SDG 8 (Decent Work and Economic Growth). This alignment dismantles the risk barriers that previously kept institutional investors at bay. We do not appeal to the traditional buyer's operational budget; we unlock the massive financial reserves of the global sustainability sector, capturing non-customer demand through absolute operational transparency and unyielding systemic discipline.

CHAPTER 33

NEURO-ERGONOMICS & AI

33.1 Cognitive Workload Optimization

Legacy agricultural automation treats the human-machine interface as a primitive afterthought. Conventional software dashboards flood operators with disorganized streams of raw telemetry, battery voltages, and unfiltered sensor alarms, forcing a single human asset to manually parse chaotic data layers while managing automated assets. This design oversight causes rapid cognitive saturation. When an enterprise scales its operations to orchestrate simultaneous multi-vector aerospace swarms, information overload ceases to be a minor fatigue issue. It becomes a critical point of systemic failure in the real sector of the economy (Mankiw, 2012). Human cognitive bandwidth is a finite thermodynamic resource. To scale an aerospace architecture across volatile global jurisdictions, a born-global firm must institutionalize advanced knowledge capabilities that actively minimize operational friction at the neurological level (Knight & Cavusgil, 2004).

We resolve this vulnerability by embedding predictive artificial intelligence directly into the core telemetry pipeline. The flight command station does not display raw, unrefined data. Instead, a localized machine-learning layer filters out baseline operating metrics, presenting the Aviation Director with clear, synthesized tactical choices only when a verified anomaly breaches hardcoded flight parameters. This execution of neuroergonomics transforms the command interface from an unyielding data repository into an adaptive analytical asset.

By systematically shielding the operator from data noise, the firm optimizes human capital processing speeds and completely insulates the operation from fatigue-induced errors. This technological intervention satisfies the core mandates of SDG 8 (Decent Work and Economic Growth). We elevate rural industrial roles by replacing exhausting, high-stress visual monitoring with high-order cognitive oversight. Human intellect is freed from digital noise, allowing a single commander to govern expanded aerospace vectors with absolute clarity and precision.

33.2 Neuro-Adaptive Value Chains

Human factors engineering must transcend the physical adjustments of control sticks and seating positions. True industrial ergonomics requires the structural alignment of digital visualization architectures with the pilot's neural processing speeds. In traditional field operations, raw physical stress and visual fatigue compromise application precision, leading to erratic flight paths, delayed emergency overrides, and systemic capital destruction. The sovereign paradigm reconfigures this dynamic from the ground up by integrating neuro-adaptive AI frameworks into the primary activities of the strategic value chain (Quintero & Sánchez, 2006). Technology development ceases to exist as an isolated corporate support function; it fuses directly with field operations to maximize asset preservation.

This neuro-adaptive synchronization serves as an ironclad operational vehicle for sustainable management and corporate responsibility (Rueda, Herrera & Rueda, 2020). The command terminal utilizes real-time gaze-tracking algorithms and cognitive load metrics to dynamically modify display layouts based on the operator's current mental fatigue profile. If the system detects an elevation in neurological friction, the interface simplifies its visual array, increases automation tolerances, and highlights critical safety vectors.

This absolute precision in human-system interaction directly enforces the targets of SDG 12 (Responsible Consumption and Production) and SDG 15 (Life on Land). By ensuring that the Aviation Director operates in a state of continuous cognitive equilibrium, the enterprise eliminates the risk of human operational delays during critical Variable Rate Application (VRA) cycles. We verifiably prevent chemical over-application, eliminate drift caused by delayed response windows, and protect the micro-structural layout of the land from operational variance. The integration of neuroergonomics and artificial intelligence establishes a highly optimized, frictionless value loop that converts human neurological efficiency into predictable, long-term capital preservation.

33.3 Multimodal Architecture Evolution

The transition of physical AI into the agricultural furrow requires the immediate abandonment of centralized, single-stream cloud architectures. Relying on remote servers to process real-time spatial and chemical telemetry introduces unacceptable latency and data ingestion bottlenecks. Up to the year 2026, the global AI landscape has undergone a massive structural paradigm shift, moving from simple reactive prompts to specialized frontier reasoning engines (GPT-5.4) and hyper-dense, native multimodal models at the edge (Gemini 3.1 Pro). This architectural evolution transforms how a born-global firm integrates its core knowledge assets (Knight & Cavusgil, 2004). The avionics suite of an optimized mid-capacity aerospace fleet operates as an autonomous cognitive ecosystem. It does not look at the field through isolated data points. It executes the simultaneous synthesis of heterogeneous, real-time sensor arrays, coordinating high-resolution computer vision, LiDAR mapping, radar, and inertial navigation data directly on the physical asset.

This edge-computing revolution solves a critical flaw in legacy corporate systems: context rot and information degradation during long-running operational windows. Traditional text or image models lose fidelity when contexts expand, causing severe cognitive drop-offs in decision-making paths. By utilizing specialized, compact edge foundation models equipped with localized context engineering, the autonomous matrix separates static operational data, such as topography and micro-structural layout matrices, from dynamic real-time telemetry.

This technological leap is a direct implementation of dynamic strategic value chains (Quintero & Sánchez, 2006). Technology development ceases to be a distant corporate cost center; it becomes an active shield for the biological asset. The vehicle's internal Black Box strategy processes millions of multimodal tokens per second without cloud dependency, allowing the onboard systems to identify crop stress, calculate aerodynamic wind variations, and adjust liquid application vectors instantly. We do not adapt our strategies to the limitations of global telecom networks. We run frontier multi-sensor intelligence natively on the silicon of our machines, ensuring that the firm's operational velocity remains unassailable.

33.4 Edge Agentic Orchestration

Managing a multi-vector autonomous fleet during complex, multi-constraint missions requires the execution of multi-agent agentic workflows. Human operators cannot be expected to orchestrate multiple independent robotic vectors simultaneously while manually adjusting to shifting environmental parameters. This manual burden triggers rapid cognitive exhaustion. By deploying multi-agent systems – where distinct, task-specific software agents plan sequences, execute multi-step tool calls, handle errors, and verify outcomes autonomously – the enterprise isolates the human factor from system entropy.

These 2026 agentic orchestration frameworks utilize parallel function calling and advanced logical planning (resembling the reasoning tracks of models like Claude Opus 4.6 or GPT-5.4 Thinking variants) to run long-duration missions without requiring human intervention for routine operational adjustments. The root agent handles macro mission objectives, delegating micro tasks to dedicated sub-agents specialized in battery optimization, chemical flow dynamics, and flight-path safety vectors.

This shift to edge-native agentic AI provides the ultimate operational structure for sustainable management and corporate social responsibility (Rueda, Herrera & Rueda, 2020). The multi-agent matrix monitors real-time environmental variables, automatically halting liquid application or adjusting drift-control parameters the exact millisecond a localized microclimate shift occurs. This high-order discipline verifiably protects the surrounding ecosystem, satisfying the rigid compliance metrics required by institutional ESG asset managers.

By automating these complex technical layers through an immutable, multi-agent ledger, the enterprise delivers on SDG 15 (Life on Land) by preventing chemical runoff, and SDG 12 (Responsible Consumption and Production) through the automated optimization of inputs.

Concurrently, this framework elevates human capital under SDG 8 (Decent Work and Economic Growth). The human asset is no longer a low-tier worker managing data noise; they become a high-level strategic commander supervising a self-correcting agentic network. We treat human intellect as a precious executive resource, leveraging advanced 2026 AI architectures to build a self-reinforcing loop of operational sovereignty and mathematical corporate stability.

CHAPTER 34
SCALING LOGIC

Linear scaling is a corporate death trap. The amateur executive believes that business expansion is a simple arithmetic problem: buy more physical vectors, hire more field operators, and secure more individual farming accounts. This asset-heavy approach sinks massive capital expenditure (CAPEX) into rapid hardware depreciation, exposing the enterprise to localized market failures, supply shocks, and systemic economic friction (Mankiw, 2012). Isolated on-farm deployment forces the firm to carry the entire weight of operational risk.

To achieve exponential market penetration, the enterprise must transcend the physical limitations of the farm furrow and operate through the lens of open innovation ecosystems. Dr. Christian A. Estay-Niculcar (2024) warns against treating corporate innovation as a closed, elite club or a superficial marketing gimmick. True organizational scaling requires a dynamic, interrelated ecosystem where the firm acts as the orchestrating node, transforming its business model from an isolated equipment provider into an institutionalized infrastructure platform.

By abandoning retail sales in favor of strategic corporate alliances, the enterprise scales through shared infrastructure. We do not sell hardware to an individual landowner; we establish joint-venture networks with global agricultural conglomerates and regional input distribution monopolies. This structure utilizes born-global capabilities to deploy knowledge-driven intellectual assets rapidly across international borders without corresponding physical asset accumulation (Knight & Cavusgil, 2004).

The real sector of the economy rewards those who control the flow of capabilities, not those who own depreciating physical structures. By embedding our proprietary control frameworks within an open, transparent business ecosystem, we shift the financial burden of hardware acquisition to corporate partners. The firm preserves its liquid reserves, captures recurring high-margin telemetry fees, and eliminates the geographic operational friction that paralyzes legacy agribusinesses.

34.2 Architecture of Transfer

True strategic scaling requires the immediate monetization of intellectual property (IP). Most emerging tech firms treat patents and trademark filings as a defensive, legal insurance expense designed to ward off litigation (Pombo Morales, 2021). This view represents severe strategic ignorance. Intellectual assets are aggressive financial instruments engineered for knowledge commercialization and technology transfer (Estay-Niculcar, 2024).

When the enterprise locks down its specialized proprietary algorithms, mechanical layout designs, and core software matrices, it creates a highly scalable, intangible asset pool that can be leveraged across commercial banking channels, multilateral subsidies, and international development funds. We do not rely on organic cash flows to build our aerospace fleets; we turn our IP portfolio into a capital-attraction engine.

Commercial banking syndicates and institutional lenders reject raw agricultural service companies due to seasonal unpredictability and structural risk. However, they aggressively finance structured technology transfer programs backed by institutionalized patents and verified risk-mitigation data. By packaging our autonomous matrix as a telemetred platform that systematically reduces chemical waste and protects long-term land values, the firm opens direct access to preferential commercial credit lines and bank guarantees.

Furthermore, global multilateral funds and government subsidies do not finance corporate growth for its own sake; they finance the verifiable resolution of public market externalities. Our architecture unifies this capital search by aligning technology commercialization with global mandates.

The verified elimination of subsoil compaction satisfies the ecological parameters of SDG 15 (Life on Land). The automated 30% reduction in chemical volumes fulfills the target metrics of SDG 12 (Responsible Consumption and Production). The transition of manual farm laborers into certified Aviation Directors directly secures the labor standards of SDG 8 (Decent Work and Economic Growth).

By utilizing Estay-Niculcar's framework of ecosystem-driven opportunity creation, we transform regulatory compliance and sustainable management into a structured mechanism for capturing extra-fiscal capital. Subsidies and green banking lines cease to be external political variables. They become non-dilutive financing streams that accelerate corporate scaling, allowing the enterprise to deploy massive aerospace swarms globally using institutional capital while maintaining absolute equity control.

34.3 Agri-Fintech Tokenization

Scaling a born-global firm requires the immediate financialization of operational data. Traditional agriculture forces landowners and service providers to wait for post-harvest liquidation to realize capital gains, creating massive cash-flow vacuums that stifle regional scaling. Commercial banks view these biological assets as high-risk, unquantifiable variables, severely restricting credit access. The sovereign enterprise shatters this bottleneck by bridging the physical furrow with the global monetary sector through Agri-Fintech architecture. The autonomous aerospace matrix is no longer deployed merely as an application tool; it functions as an indisputable auditing mechanism.

Through the edge-native multimodal networks discussed previously, the swarm captures real-time, unforgeable telemetry regarding crop health, precise chemical reduction, and yield trajectories. Instead of storing this data passively, the enterprise leverages it to execute Biological Yield Tokenization. We convert verified agronomic data into dynamic smart contracts.

This ecosystem orchestration completely alters the macroeconomic value chain (Quintero & Sánchez, 2006). Banks and multilateral subsidy funds do not finance the traditional, high-risk farmer; they finance the cryptographically verified data token generated by our autonomous fleet. This unlocks immediate capital liquidity. Associations of independent growers, previously marginalized by corporate banking, can now access premium credit lines because the AgTech commander underwrites their operational risk with military-grade aerospace telemetry. The enterprise scales exponentially, extracting high-margin transactional fees from the financial sector while simultaneously satisfying the transparent reporting mandates of institutional ESG portfolios.

34.4 Venture Ecosystem Architecture

The final phase of exponential scaling abandons the archaic strategy of organic, linear growth in favor of Corporate Venture Capital (CVC) and strategic Mergers and Acquisitions (M&A). Attempting to conquer the global market by purchasing individual localized contracts is an exercise in corporate exhaustion. According to the foundational parameters of innovation ecosystems (Estay-Niculcar, 2024), true dominance requires the enterprise to position itself as the orchestrating hub of a much larger, interrelated network.

The enterprise scales by structuring joint ventures and executing targeted acquisitions of adjacent technologies. These include localized soil-sensor startups, proprietary biological nutrient manufacturers, or regional cooperative associations. By acquiring distressed or undercapitalized regional operators, the born-global firm immediately replaces their obsolete manual infrastructure with our automated, Black Box flight architecture (Knight & Cavusgil, 2004). We acquire market share at a severe discount and instantly force it into high-margin profitability through our thermodynamic efficiency.

This architecture acts as an aggressive barrier to entry. Competitors attempting to enter the market face an impenetrable ecosystem where the autonomous fleet, the data analytics, and the banking credit lines are inextricably linked and controlled by a single sovereign entity. By institutionalizing this ecosystem, the firm enforces the highest levels of corporate social responsibility. Bringing fragmented, non-compliant regional operators into our disciplined matrix immediately halts their ecological destruction, enforcing SDG 15 (Life on Land) and SDG 12 (Responsible Consumption and Production) across acquired territories. We do not just scale a business; we scale a mandatory protocol for biological preservation and absolute financial hegemony.

CHAPTER 35

FOOD SOVEREIGNTY

35.1 Geopolitical Yield Weaponization

The global food supply is the ultimate kinetic weapon. Traditional macroeconomic models treat agricultural outputs as passive commodities, perpetually vulnerable to the chaotic price fluctuations of international trade corridors and fiat currency manipulation (Mankiw, 2012). This is a fatal strategic vulnerability. When a nation or a corporate entity relies on imported agrochemicals, foreign cloud-based telemetry, and distant logistical hubs to feed its population, it surrenders its sovereign autonomy. True food sovereignty is never achieved through passive government policy or superficial ESG campaigns. It must be engineered. By deploying decentralized, autonomous aerospace swarms, the enterprise actively decouples localized biological production from global supply chain friction.

The Agri-Fintech tokenization of biological yield must transcend its role as a mere financial instrument for venture capitalists. It must become a mechanism of geopolitical defense. We do not just monetize data; we leverage thermodynamic efficiency. When an enterprise controls the exact micro-inputs of liquid applications on the biological substrate and encrypts that telemetry within a strict Black Box architecture, it insulates the harvest from external state-sponsored sabotage, inflationary shocks, and trade embargoes. We bypass the external sector entirely. Food sovereignty is the mathematical elimination of foreign dependency. Whoever controls the edge-native cognitive algorithms governing the agricultural furrow dictates the geopolitical leverage of the region.

35.2 Autarkic Infrastructure Matrix

Autarky is the foundation of structural dominance. Legacy agriculture operates on an extractive model that continuously bleeds operational capital to external heavy-hardware manufacturers and global chemical monopolies. A born-global firm dismantles this dependency through the strategic integration of proprietary aerospace architectures, utilizing optimized hexagonal deployment matrices (Knight & Cavusgil, 2004). The physical architecture of the flight swarm is not merely a retail delivery mechanism for liquid spraying; it is a mobile, self-sustaining infrastructure grid.

By fusing edge-native artificial intelligence with biological co-investment frameworks, the enterprise establishes absolute localized autarky. We eliminate the structural reliance on heavy terrestrial tractors that neutralize the macroporosity of the soil. We eradicate dependency on vulnerable foreign telecom networks by processing all multispectral data locally at the node. This closed-loop system forces the macroeconomic environment to adapt to the enterprise, rather than the enterprise bleeding capital to react to external sector shocks. When the physical means of liquid application, the proprietary cognitive flight software, and the financial tokenization of the yield are entirely consolidated under a single command structure, the enterprise transcends the status of an AgTech service provider. It becomes an unassailable, sovereign entity within the agricultural domain, dictating the terms of survival to the broader market.

35.3 Chemical Cartel Disruption

The traditional agricultural supply chain is economically subjugated by global agrochemical cartels. Legacy farming practices operate on systemic thermodynamic inefficiency, deliberately over-saturating the biological substrate to compensate for the analogue inaccuracy of heavy terrestrial machinery. This is not agronomic science; it is forced capital extraction. By deploying an autonomous aerospace matrix exclusively engineered for ultra-precision liquid spraying, the sovereign enterprise executes a targeted disruption of these chemical monopolies.

Variable Rate Application (VRA) algorithms, shielded within the unyielding Black Box architecture, slash liquid input volumes by up to 30%. We do not merely reduce operational expenses; we systematically starve the legacy chemical supply chain of its inflated margins. This optimization of the Total Cost of Ownership (TCO) fundamentally shifts the macroeconomic balance of power. The landowner is no longer a captive, high-volume consumer of chemical commodities, but a strategic co-investor in a mathematically optimized biological asset. The enterprise captures the massive economic surplus previously lost to chemical drift and waste, converting it directly into high-velocity corporate capital. Food sovereignty demands that the financial leverage of the harvest remains with the intelligence governing the node, not with the foreign manufacturers of the raw inputs.

35.4 Regulatory Capture Metrics

Sovereignty is fragile if it remains vulnerable to hostile state legislation or bureaucratic interference. A born-global firm does not wait for municipal agencies to dictate operational boundaries; it actively engineers regulatory capture. Conventional competitors view institutional ESG compliance and the Sustainable Development Goals (SDGs) as a superficial public relations tax. The AgTech commander leverages them.

By institutionalizing the unforgeable telemetry of our liquid spraying operations within a dynamic innovation ecosystem (Estay-Niculcar, 2024), we establish a new, insurmountable industry standard for ecological protection. When the enterprise proves mathematically that its Black Box architecture verifiably enforces SDG 12 (Responsible Consumption and Production) and SDG 15 (Life on Land), completely eliminating toxic runoff and subsoil compaction, we deploy this data directly into aggressive aerospace lobbying efforts.

The strategic objective is to force legislative bodies to mandate these exact telemetric compliance levels for all agricultural operations. Traditional operators, burdened by obsolete ground tractors and lacking edge-native cognitive AI, are instantly rendered illegal or economically unviable by the new regulatory thresholds. We do not compete with legacy contractors; we legislate them out of existence. The sovereign enterprise writes the laws of the agricultural furrow, transforming sustainable compliance into a state-enforced corporate monopoly.

CHAPTER 36

MILITARY STRATEGY APPLIED

36.1 Tactical Asset Weaponization

The conventional agricultural sector operates under a fatal misconception, treating physical infrastructure and intellectual property as static, defensive garrisons. Traditional agribusinesses deploy hardware reactively, responding to seasonal demands while passively defending their market share against price undercutting. This is a strategy of attrition, guaranteeing margin erosion and systemic fatigue. The sovereign AgTech commander discards this civilian mindset, applying strict military doctrine to market penetration. Intellectual property is not a bureaucratic shield; it is a kinetic weapon engineered to preemptively neutralize the opposition.

By legally locking down the proprietary architecture of our aerospace fleets, we execute a tactical blockade on the market. We do not disclose our specific aerodynamic configurations, proprietary structural dampening modules, or advanced operational power matrices merely to prevent industrial theft. We register them to dictate the absolute boundaries of the combat theater. Competitors attempting to replicate the thermodynamic efficiency required for high-capacity precision agriculture will not face a standard commercial rivalry; they will face immediate, decisive international litigation. By weaponizing our intellectual assets, we transform raw engineering superiority into an unassailable legal perimeter. The enterprise commands the airspace long before the autonomous swarm ever leaves the ground.

36.2 Asymmetric Swarm Dominance

In modern tactical operations, asymmetric advantages dictate the survival of the unit. The legacy agricultural market relies on symmetric brute force – heavy terrestrial tractors and generic, quadcopter drone frames engaging in a localized war of price attrition. We shatter this symmetry. The sovereign enterprise deploys a highly specialized, hexagonal aerospace matrix, explicitly engineered for the thermodynamic demands of ultra-low volume liquid spraying.

This hexagonal geometry is the physical manifestation of asymmetric warfare. It maximizes structural integrity and payload stability while minimizing energy consumption, rendering the generic hardware of conventional contractors mathematically obsolete. When you pair this physical superiority with a strict Black Box operational strategy, the advantage becomes impenetrable.

The edge-native cognitive algorithms governing the liquid application vectors are permanently encrypted, hidden from foreign cloud networks and industrial espionage. We execute surgical strikes on biological targets, slashing chemical waste by 30% and neutralizing phytosanitary threats with zero collateral damage to the subsoil macroporosity. The competition is left systemic optimizationing an invisible enemy. They cannot reverse-engineer the flight telemetry, and they cannot legally replicate the hexagonal hardware. This is the essence of applied military strategy: you do not compete with the enemy; you systematically dismantle their capacity to engage.

36.3 Licensing Biological Profit

Amateur strategists accumulate physical liabilities. The fatal flaw of traditional AgTech expansion is the reliance on linear scaling: purchasing more hardware, hiring more manual labor, and chasing individual localized contracts. This model absorbs massive Capital Expenditure (CAPEX) and exposes the firm to rapid asset depreciation. The sovereign enterprise completely abandons this operational quicksand. By weaponizing our intellectual property, we shift the corporate structure from a vulnerable service provider to an apex ecosystem licensor.

We do not scale by deploying our own capital to manufacture infinite fleets. We scale by licensing the proprietary closed-loop ecosystem. Regional agricultural cooperatives and global agribusinesses possess the localized capital and the biological assets, but they lack the technological sovereignty to optimize them. We license access to our patented hexagonal matrices, our aerospace-grade polymer configurations, and our encrypted Black Box navigation software. In this framework, the enterprise extracts high-margin, recurring royalty streams and captures all yield telemetry, while the licensee absorbs the physical hardware costs. This is the ultimate execution of financial hegemony. We do not sell the physical act of liquid spraying; we license the guaranteed, mathematically verified optimization of biological profit.

36.4 The Cognitive Commander

Total operational dominance cannot be achieved if the human interface remains an uncalibrated variable. Traditional agriculture systematically optimizes its human capital, relying on brute physical endurance and exposing laborers to neurotoxic chemical drift. This archaic model breeds extreme physical fatigue, emotional burnout, and catastrophic operational entropy. A sovereign aerospace matrix requires an entirely different psychological architecture. To command a high-velocity, decentralized fleet, the human asset must be elevated from a biological tractor into a precision Aviation Director.

This elevation requires rigorous psychological resilience and advanced emotional regulation. When operating at the nexus of edge-native artificial intelligence and biological volatility, panic or cognitive saturation results in immediate capital destruction. The commander's mind must mirror the thermodynamic efficiency of the machine. By eliminating exhausting physical toil and integrating neuro-adaptive command interfaces, the enterprise cultivates a culture of cold, cognitive discipline.

The operator does not sweat in the furrow; they regulate complex telemetric data streams, adjusting liquid application vectors and mitigating geopolitical supply risks in real-time. This intellectual elevation fulfills the most stringent mandates of sustainable corporate responsibility, permanently dignifying the rural workforce. We do not merely engineer advanced machines. We engineer the psychological fortitude of the operator, forging a human-machine symbiosis that no traditional competitor can ever disrupt.

CHAPTER 37

OPERATIONAL HOMEOSTASIS

37.1 Dynamic Equilibrium

In the chaotic theater of global agribusiness, traditional operations oscillate wildly between peak harvests and catastrophic financial contractions. This volatility is a symptom of structural fragility. The sovereign enterprise eliminates this vulnerability by engineering Operational Homeostasis: the mathematical certainty that the organization will maintain absolute internal equilibrium despite violent external market disruptions. Homeostasis is not merely a static accounting metric or a theoretical goal; it is a dynamic, living ecosystem where cash flow velocity, capital structure, and operational deployment are perfectly synchronized.

Traditional parcel farmers and legacy agribusinesses rely on reactive survival tactics. When market prices drop or supply chains fracture, they descend into operational chaos, making hasty, emotionally driven decisions that erode stakeholder confidence. The AgTech commander constructs a framework of absolute stability. We establish a state of financial homeostasis where automated risk management, predictive data analytics, and diversified revenue streams act as structural shock absorbers. By harmonizing internal operational efficiency with the volatile external demands of the macroeconomic environment, the enterprise becomes immune to systemic panic. Investors do not merely fund our technology; they invest in our engineered stability. We do not react to the chaos of the market; we absorb its volatility and mathematically convert it into sustained, predictable growth.

37.2 Antifragile Architecture

True financial and operational homeostasis demands the total eradication of reactive management. As W. Edwards Deming established in his Total Quality Management (TQM) frameworks, any system that relies on mass inspection or post-failure reaction is a system hemorrhaging capital. In the context of high-capacity precision agriculture, waiting for a high-discharge lithium-polymer battery to catastrophically swell or an Electronic Speed Controller (ESC) to melt before initiating repairs is a symptom of sheer managerial incompetence.

To enforce absolute stability, the enterprise aggressively models its Mean Time Between Failures (MTBF). The physics of heavy-lift aerospace platforms mandate that thermal degradation is constant; hovering a 40-liter liquid payload over a scorching field subjects the propulsion matrices to extreme thermodynamic stress. The sovereign operation neutralizes this through Predictive Acoustic and Vibrational Analytics. Edge-native cognitive algorithms continuously analyze the micro-vibrations of the brushless motors. An anomaly of merely 0.5 Hz serves as an immediate, early indicator of bearing fatigue, flagging the component for replacement days before a critical failure can occur.

This resilience is strictly codified in the Total Cost of Ownership (TCO) equation. If operational uptime drops below 92% during a critical phytosanitary intervention window, the cost of the hardware becomes statistically irrelevant because the biological asset, the harvest, is lost. Therefore, homeostasis requires the structural integration of Nassim Taleb's concept of Antifragility. The logistics chain must not merely withstand operational shocks; it must improve from them. By maintaining a decentralized, hyper-localized inventory buffer of Line Replaceable Units (LRUs) and executing military-grade thermal management protocols (in accordance with the NTA 8133 standard), the fleet absorbs component failures without ever disrupting the sortie generation rate. The operation breathes, adapts, and continues to spray, completely decoupled from the chaotic entropy of the physical environment.

37.3 Algorithmic Capital Routing

Retrospective financial analysis is corporate autopsy. Conventional business directors rely on end-of-quarter balance sheets to identify operational bleeding, responding to market volatility months after the capital has already evaporated. This delayed reaction guarantees systemic imbalance. The sovereign enterprise executes absolute financial homeostasis through real-time, algorithmic capital routing. The autonomous aerospace matrix is not merely a localized agricultural tool; it functions as a decentralized, high-velocity financial auditing node.

Powered by edge-native artificial intelligence, the swarm continuously ingests multispectral telemetry, chemical flow rates, and meteorological volatility. If the onboard cognitive matrix detects a sudden micro-climate shift, such as a thermal inversion that degrades liquid application efficacy, it does not wait for manual managerial approval. It autonomously halts the sortie, recalibrates the Variable Rate Application (VRA) algorithms, or grounds the fleet entirely. This is the absolute weaponization of data analytics.

We eliminate human emotional delay from the financial equation. Every drop of chemical saved and every kilowatt of battery life preserved is instantly quantified and locked into the corporate ledger. The system self-corrects at the speed of silicon. By fusing operational execution directly to real-time data analytics, the enterprise guarantees that operational expenditure (OPEX) remains mathematically bound to the biological yield projection. Financial stability is no longer a goal; it is a hardcoded algorithmic certainty.

37.4 Neuro-Resilient Command

Chaos thrives in the gap between human perception and mechanical execution. In traditional agribusiness, rapid external shocks – supply chain fractures, sudden phytosanitary outbreaks, or macroeconomic contractions – trigger immediate operational panic. Emotional decision-making under stress optimizes carefully constructed financial ecosystems. To achieve true organizational homeostasis, the enterprise must rigorously engineer the psychological architecture of its leadership. We cultivate Neuro-Resilient Command.

The Aviation Director governing the swarm does not interact with raw, chaotic data streams. The Black Box interface acts as a cognitive shield, filtering out environmental noise and presenting only synthesized, high-order tactical vectors. This neuro-adaptive synchronization prevents cognitive saturation. By isolating the human operator from the mechanical entropy of the physical field, the enterprise eradicates the stress-induced paralysis that plagues legacy contractors. The commander evaluates risk with cold, clinical precision, executing strategic pivots without hesitation.

This psychological homeostasis permeates the entire corporate structure. Leadership under this doctrine does not fear market turbulence; it harnesses it. The organization operates with the cold discipline of a military unit, maintaining unwavering focus on Total Cost of Ownership (TCO) optimization and biological asset protection. When the human mind mirrors the thermodynamic stability of the hexagonal flight matrix, failure becomes structurally impossible.

CHAPTER 38

CYBER-RESILIENCE

38.1 Algorithmic Autarky

The civilian AgTech sector operates under a fatal illusion regarding data management. Traditional operators willingly outsource their agricultural telemetry to third-party, globalized cloud infrastructures, treating their most critical operational intelligence as an exportable commodity. This surrender of data sovereignty is a structural vulnerability. In precision agriculture, data is not merely a record of past events; it is the predictive blueprint of the biological asset. When a farm relies on foreign servers and external software architectures to process crop health metrics and soil conditions, it relinquishes control of its primary economic engine. A service outage, a foreign regulatory shift, or a targeted cyber breach can instantly paralyze the operation.

The sovereign enterprise rejects cloud dependency in favor of Algorithmic Autarky. We establish absolute data sovereignty by shifting the computational burden from distant servers directly to the physical furrow. By integrating localized edge-native supercomputing within the hexagonal drone matrix, the swarm processes hyperspectral imagery, thermal gradients, and LiDAR topography in real-time, completely independent of global telecommunication networks.

This is the convergence of cyber resilience and financial hegemony. The enterprise retains absolute cryptographic ownership of the agronomic data, insulating the operation from external exploitation. We do not transmit the granular biological data of our high-value assets across vulnerable international borders. We process it on the silicon of our own machines, storing the intelligence in localized, encrypted data nodes. This uncompromising control mechanism ensures that the intellectual property generated by the soil remains the exclusive financial leverage of the Commander, transforming data sovereignty from a theoretical legal concept into an impenetrable operational fortress.

38.2 Kinetic Sabotage

Cybersecurity in advanced aerospace agriculture cannot be treated as a traditional IT function. In the context of autonomous heavy-lift swarms, a cyberattack is not a data leak; it is a kinetic strike on a biological target. Legacy agribusinesses fail to comprehend the weaponization of the electromagnetic spectrum. A hostile entity, whether a corporate rival or a state-sponsored saboteur, deploying GNSS (Global Navigation Satellite System) spoofing or localized signal jamming can hijack civilian drone coordinates. The result is catastrophic physical entropy: aerospace assets crashing into the terrain or drifting to release chemical payloads over restricted ecological zones.

Furthermore, the ultimate threat is algorithmic poisoning. If an unauthorized actor breaches the command terminal and overrides the Variable Rate Application (VRA) algorithms, they could command the drone's micro-pumps to deploy 100% concentrated phytosanitary chemicals. This would instantly incinerate the cellular integrity of the crop canopy, annihilate the delicate microbiological networks of the subsoil, and permanently sterilize the land's macroporosity. Cyber resilience is, therefore, the absolute defensive perimeter around the plant's biology.

To neutralize these threats, the physical engineering of the drone must act as an ironclad biological shield. The sovereign hexagonal matrix utilizes multimodal Visual Odometry and advanced Inertial Navigation Systems (INS). If the external GPS signal is jammed, the Black Box architecture instantly severs the connection, allowing the swarm to navigate blindly using internal gyroscopic physics and localized optical flow sensors. More importantly, the fluid dynamics of the liquid spraying system are hardcoded with physical, thermodynamic fail-safes. Even if the software matrix is entirely compromised by a hostile actor, the mechanical aperture of the nozzles physically restricts the chemical flow rate, preventing biological over-saturation. We mathematically eliminate the possibility of total agronomic failure, ensuring the harvest remains untouchable by external cyber-kinetic entropy.

38.3 Immutable Ecosystem Protocols

The concept of open collaborative ecosystems is a civilian vulnerability. Legacy agricultural technology attempts to integrate fragmented hardware, standard commercial software, and local farm Wi-Fi into a single operational network. This creates an infinitely expanding attack surface where a single compromised regional sensor can infect the entire corporate database. The sovereign enterprise does not "collaborate" on cybersecurity; it dictates it. When engaging in strategic Mergers and Acquisitions (M&A) or licensing our technology to regional cooperatives, we do not integrate our Black Box architecture into their existing infrastructure. We obliterate their legacy networks and impose our immutable cryptographic protocols.

This is the execution of Zero-Trust Architecture in the agricultural furrow. We assume that every external network, every vendor, and every regional internet service provider is already compromised. To maintain algorithmic autarky, the hexagonal aerospace swarm deploys its own encrypted, decentralized mesh network over the biological asset. If a newly acquired regional farm wishes to operate under the sovereign banner, its obsolete digital infrastructure is systematically purged. Every data transmission, from the Variable Rate Application (VRA) telemetry to the supply chain logistics, is forced through our proprietary encryption tunnels. By imposing our cyber-laws upon our partners, we ensure that the enterprise scales globally without ever diluting its defensive perimeter. Collaboration is permitted only through absolute cryptographic compliance.

38.4 Autonomous Forensic Recovery

Even with military-grade shielding, absolute invincibility is a mathematical impossibility in a physical environment. A sovereign aerospace matrix does not rely on the naive assumption that a breach will never occur; it relies on the absolute certainty of instantaneous recovery. When a targeted cyber-kinetic attack or extreme electromagnetic interference successfully corrupts a single drone within the flight swarm, traditional operators ground the entire fleet, hemorrhaging capital while awaiting manual IT intervention.

The AgTech commander engineers the swarm to execute algorithmic amputation. If the edge-native cognitive matrix detects a logic fault or an unauthorized command attempt within a specific drone, the compromised unit is instantly severed from the mesh network. The Black Box executes an autonomous cryptographic purge, permanently deleting the proprietary flight algorithms and VRA chemical configurations from the infected node's local memory, neutralizing any threat of industrial espionage.

Simultaneously, the remaining drones in the swarm recalculate the aerodynamic vectors and redistribute the phytosanitary payload to cover the amputated unit's sector. The sortie continues uninterrupted. The biological asset remains protected, and the Total Cost of Ownership (TCO) thermodynamic efficiency is preserved. Once the mission is completed, the system generates an autonomous forensic audit, analyzing the attack vector without human bias. We do not halt operations to investigate a failure; we isolate the casualty, finish the biological intervention, and leverage the forensic data to harden the swarm for the next deployment.

CHAPTER 39

THE STRATEGY OF THE SOUL

39.1 The Physics of Operational Resonance

The concept of "human energy" is often relegated to the realm of intuition, yet in complex systems engineering, it is a quantifiable physical reality. Humans are not static observers of the production process; we are biological oscillators. Every operational intent, driven by technical focus, manifests as a high-frequency cognitive state, a phenomenon of neural entrainment. In the sovereign enterprise, this is not a subjective quality; it is Operational Resonance.

Systems theory dictates that when the human operator's cognitive frequency aligns with the mechanical output of the aerospace swarm, the result is the elimination of "informational lag." This is the state of Ω (Omega) in our sovereign identity equation. When the operator is fully synchronized with the thermodynamic reality of the drone, their decision-making process functions at the speed of the machine's own processing architecture. This is not "spiritual" energy; it is the physical synchronization of biological neural networks with electronic command protocols.

39.2 The Law of Thermodynamic Reciprocity

In the Newtonian framework, every action produces an equal and opposite reaction. In the sovereign enterprise, this manifests as the Law of Thermodynamic Reciprocity: the intensity of disciplined, intent-driven labor invested into the biological substrate returns as a higher frequency of system efficiency. This is why "passion", the relentless drive for technical perfection, is a tangible driver of the profit equation. It reduces entropy. An operation managed with high-frequency focus experiences lower mechanical friction, fewer data errors, and tighter flight-path tolerances. The system, quite literally, vibrates at a higher level of efficiency because the human component is operating in perfect harmonic resonance with the mechanical and environmental variables.

This is the secret of the Sovereign Endgame: Hegemony is achieved not just through superior silicon, but through the seamless integration of human intent into the thermodynamic loop of the machine. The enterprise that masters this frequency – this precise alignment of mind, matter, and methodology – attains a level of operational fluidity that renders competition irrelevant.

39.3 Predictive Yield Oracles

Legacy agriculture utilizes crop yield prediction models as mere logistical tools, estimating harvest volumes to schedule transport trucks and warehouse capacity. This is a profound underutilization of data. The sovereign AgTech commander does not predict the yield to manage logistics; they predict the yield to manipulate the macroeconomic market. By fusing drone-assisted genetic monitoring with edge-native artificial intelligence, the aerospace matrix is transformed into a highly classified, algorithmic financial oracle.

While conventional operators wait for the harvest to reveal their financial fate, the sovereign enterprise maps the exact biological trajectory of the high-value specialty asset months in advance. The swarm processes multi-layered telemetry – soil carbon retention, cellular stress markers, and biopesticide efficacy – generating an infallible mathematical projection of the final biological output. This data is permanently encrypted within the Black Box architecture. We possess absolute asymmetric information. By knowing the exact volume, quality, and metabolic health of the harvest long before it is extracted from the soil, the enterprise completely decouples its revenue from the chaotic fluctuations of the spot market. The competitors are playing an agricultural game of chance, hoping the weather and pests spare their crops. The sovereign enterprise plays a rigorous game of financial physics, where the biological yield is merely the underlying asset for a much larger, mathematically guaranteed economic victory.

39.4 The Biological Hedge

When an enterprise possesses an infallible predictive oracle, it ceases to be a traditional agricultural service provider. It becomes a biological hedge fund. This is the ultimate synthesis of the Commander's Soul: merging agronomic reality with advanced quantitative trading. The financialization of the crop is executed long before the physical harvest.

Armed with precise, drone-generated yield models, the corporate leadership executes aggressive arbitrage in the commodities and futures markets. If the aerospace swarm detects a subtle, regional phytosanitary vulnerability or a micro-drought that will devastate competitors' yields, the enterprise does not merely protect its own assets. It actively shorts the market. We leverage our proprietary data to place calculated financial bets against the blind spots of the legacy agricultural sector.

Conversely, our verified, algorithmically protected yield projections allow us to sell forward contracts at premium valuations, backed by the cryptographic certainty of our autonomous matrix. The competitors are playing an agricultural game of chance, hoping the weather and pests spare their crops. The sovereign enterprise plays a rigorous game of financial physics, where the biological yield is merely the underlying asset for a much larger, mathematically guaranteed economic victory.

Operational success within the sovereign enterprise is not defined by chance, but by Operational Resonance, a quantifiable state of Biomechanical Entrainment . This is the harmonic synchronization between the operator's cognitive frequency and the system's execution architecture, where the human operator and the aerospace swarm function as a single, entropic-resistant organism. This resonance is the foundation upon which the following predictive and biological strategies are built.

CHAPTER 40

REGULATORY ARBITRAGE

40.1 Leveraged Compliance

The agricultural sector treats the Sustainable Development Goals (SDGs) and environmental regulations as punitive taxation, a bureaucratic burden to be minimized through public relations. This is a civilian mindset. The sovereign AgTech commander views the EPA (Environmental Protection Agency), the FAA (Federal Aviation Administration), and the United Nations SDGs not as obstacles, but as the ultimate instruments of corporate warfare. We do not merely comply with environmental legislation; we leverage it to outlaw the competition.

Traditional terrestrial tractors and amateur operators live in a state of continuous regulatory vulnerability. They rely on massive chemical volume. This inevitably causes toxic runoff and hazardous material drift. The sovereign enterprise deploys its aerospace matrix to execute targeted applications with mathematical exactitude. We utilize native edge artificial intelligence to guarantee absolute precision. We generate cryptographic and immutable proof of compliance with SDG 12 and SDG 15.

This telemetric proof is locked in our Black Box. We then execute aggressive regulatory capture. The enterprise approaches environmental regulators. We do not ask for operational permits. We lobby for stricter HAZMAT mandates. We present our proprietary fluid kinetics and autonomous swarm orchestration as the new baseline for ecological safety. We force regulators to adopt our patented thermodynamic efficiency as the legal standard for agricultural spraying. This action instantly renders legacy machinery and amateur fleets illegal. We do not compete on price. We use the sustainability mandates of the government to permanently revoke the operational rights of our competitors.

40.2 Jurisdictional Asymmetry

Regulatory arbitrage is the strategic exploitation of the friction between intersecting governing bodies. In the airspace above the furrow, two massive bureaucratic entities collide: the FAA governs the kinetic reality of the flight, while the EPA governs the biological reality of the chemical payload. Amateur operators and regional startups are crushed in the jurisdiction gap between Part 107 aviation constraints and stringent HAZMAT transport laws. They cannot scale because they treat regulatory compliance as an administrative hurdle rather than an engineering problem.

The sovereign enterprise transcends this gridlock through systemic asymmetry. We do not ask the FAA for standard Beyond Visual Line-of-Sight (BVLOS) waivers for isolated drones. We submit petabytes of localized, encrypted Visual Odometry and Inertial Navigation System (INS) data, proving that our autonomous swarms operate with a multi-layered, cyber-resilient fail-safe architecture. We demonstrate that in the event of a cyber-kinetic attack or GNSS spoofing, the swarm mathematically cannot drop a HAZMAT payload outside the designated coordinates.

By satisfying the EPA's deepest environmental fears through the FAA's rigid aerospace safety protocols, the enterprise secures exclusive, sweeping operational authorizations that are denied to the rest of the market. This creates an impenetrable legal monopoly. The competition is trapped systemic optimizationing for basic line-of-sight permits, while the sovereign commander orchestrates autonomous, multi-vector swarms across vast territories. Regulatory arbitrage is not about finding loopholes; it is about engineering a technological standard so advanced that the regulatory bodies have no choice but to grant you absolute airspace hegemony.

40.3 Data-Driven Lobbying

Traditional corporate lobbying relies on political capital, financial influence, and public relations rhetoric. It is a fragile, outdated mechanism. The sovereign AgTech enterprise replaces political friction with undeniable, cryptographic mathematics. When engaging with international agricultural ministries or federal agencies like the FAA and EPA, we do not present theoretical whitepapers or hollow sustainability pledges. We present petabytes of immutable, Black Box telemetry.

We lay the raw data on the regulatory table: the exact fluid dynamics of our proprietary Variable Rate Application (VRA) algorithms, the thermodynamic stability of the hexagonal aerospace matrix, and the absolute elimination of chemical drift. We demonstrate that our elastomeric-damped propulsion systems and edge-native cognitive algorithms achieve a level of phytosanitary precision that legacy equipment simply cannot replicate.

This is the execution of data-driven lobbying. We align our technological superiority directly with the government's desperate need to fulfill global mandates like SDG 12 (Responsible Consumption and Production). We do not ask for leniency. We demand that regulators elevate the national legal standards to match our exact operational metrics. We lobby lawmakers to mandate zero-drift tolerances and strict cryptographic reporting for all aerial hazardous material (HAZMAT) applications. By forcing the state to adopt our patented engineering thresholds as the baseline for legal operation, we effectively outlaw legacy machinery. The government becomes the enforcer of our technological monopoly, mistaking our aggressive corporate blockade for ecological salvation.

40.4 The Compliance Moat

Empathy in the agricultural sector is rarely directed where it belongs. Traditional markets shed tears for the obsolescence of outdated machinery while ignoring the silent destruction of the biological asset.

True empathy demands the absolute protection of the specialty asset's biological integrity and the economic dignity of the localized operator. For decades, regional parcel farmers have been crushed under the dual weight of chemical cartel pricing and increasingly draconian state regulations. The sovereign enterprise constructs a fortress to shield them.

This fortress is the Compliance Moat. By establishing an operational standard so advanced that it seamlessly navigates the jurisdictional asymmetry between aviation safety and environmental law, we create a sanctuary for the biological yield. When a regional specialty enterprise operates under our licensed ecosystem, they inherit our regulatory immunity. They are no longer vulnerable to sudden EPA audits or FAA grounding orders, because the autonomous swarm mathematically guarantees their compliance before the liquid payload even leaves the nozzle.

This moat is decisive to competitors. An amateur startup attempting to enter the precision agriculture space is immediately paralyzed. They are forced to burn massive capital merely trying to achieve the legal certifications that our architecture generates automatically. They cannot afford the complex regulatory compliance, the proprietary high-stress material structural requirements, or the closed-loop cyber resilience required to meet the new, elevated legal standards. We leverage the law to protect the soil, simultaneously creating a barrier to entry so massive that the only viable option for the market is total submission to our ecosystem. The regulatory theater is conquered. The board is set for absolute geopolitical and financial expansion.

CHAPTER 41

COGNITIVE ARCHITECTURE

41.1 The Ethical Algorithm

The deployment of advanced aerospace technology inherently triggers a geopolitical paradox. As drone capabilities expand, they inevitably collide with the ethical risks of mass surveillance, privacy infringement, and civil liberty violations. Legacy agribusinesses and amateur operators are entirely unequipped to handle the moral and legal blowback of capturing unauthorized data across rural territories. When a civilian drone accidentally records neighboring properties or critical infrastructure, it creates a massive geopolitical and financial liability. The sovereign enterprise does not rely on fragile corporate policies or public relations apologies to mitigate this risk. We solve the ethical dilemma through hardcoded cognitive architecture.

Our aerospace swarms are engineered with an uncompromising operational mandate: the absolute separation of agronomic telemetry from geopolitical surveillance. The edge-native artificial intelligence governing the swarm utilizes localized neural processing to instantly filter and permanently delete any visual or spatial data that does not pertain to the biological asset. Human figures, neighboring vehicles, and adjacent properties are algorithmically redacted at the edge, before the data is ever encrypted into the Black Box.

We do not debate surveillance ethics with regulators or the media; we mathematically eliminate the capacity for espionage. By designing a cognitive architecture that is functionally blind to everything except the thermodynamic and biological reality of the crop canopy, the enterprise neutralizes public fear and media-driven panic. We convert ethical compliance from a bureaucratic vulnerability into a cryptographic certainty, ensuring our investors and regional partners operate under an impenetrable shield of social governance.

41.2 Ecological Silence

In the pursuit of maximum yield, the agricultural industry has historically operated with deafening brutality. Heavy terrestrial machinery crushes the subsoil, while primitive agricultural drones generate acoustic profiles and thermal exhaust that severely disrupt local wildlife, breeding patterns, and the surrounding ecological equilibrium. The provided literature correctly identifies that uncalibrated drone deployment poses severe biological risks to natural habitats. The sovereign commander recognizes that disrupting the local ecosystem is not just an environmental failure; it is a rapid destruction of biological capital.

To achieve true operational homeostasis, the cognitive architecture of the autonomous matrix must extend beyond visual telemetry to master acoustic and environmental fluid dynamics. We execute Ecological Silence.

The swarm's central processing node continuously calculates the optimal aerodynamic vectors required to minimize rotor RPMs and dampen acoustic signatures. Proprietary vibration-dampening modules and high-stress structural matrices absorb the micro-vibrations that traditional deployment assets broadcast into the environment. The cognitive architecture dictates flight paths that actively avoid sensitive ecological zones, adjusting altitude and speed in real-time to prevent the disruption of local fauna.

We protect the farmer's biological asset without annihilating the surrounding ecosystem. This is not passive environmentalism; it is aggressive, calculated stewardship. By proving that our heavy-lift aerospace matrices operate with near-zero ecological friction, we completely dismantle the environmental arguments utilized by hostile regulators and competitors. The cognitive architecture proves that absolute technological supremacy and absolute ecological preservation are the exact same mathematical equation.

41.3 Algorithmic Accountability

In the traditional geopolitical theater, accountability is a bureaucratic afterthought. It relies on fragile human oversight, subjective post-incident reports, and armies of compliance officers scrambling to construct a legal defense after a failure has already occurred. When legacy operators breach an environmental threshold or violate a privacy mandate, they offer public apologies and absorb crippling regulatory fines. The sovereign AgTech commander eradicates this vulnerability. We do not rely on legal defense; we rely on cryptographic proof. We replace human compliance with Algorithmic Accountability.

Every micro-decision executed by the swarm's cognitive architecture is hashed into a localized, immutable cryptographic ledger. From the exact millisecond a Variable Rate Application (VRA) nozzle opens, to the specific aerodynamic pivot required to maintain ecological silence, the drone writes its own unforgeable operational history. If a regulatory body or an institutional ESG investor questions the ethical or environmental validity of a sortie, we do not engage in debate. We deliver the cryptographic hash.

This level of transparency is decisive to our competitors. By proving mathematical compliance with both aerospace safety and biological preservation, we expose the catastrophic opacity of legacy agricultural practices. The cognitive architecture turns the drone into an infallible auditor of its own actions. We neutralize geopolitical blowback and institutional skepticism not with public relations, but with absolute data sovereignty. When accountability is hardcoded into the silicon, the enterprise becomes legally and ethically unassailable.

41.4 The Governance Engine

A machine that calculates, regulates, and censors its own ethical parameters is no longer merely a piece of agricultural hardware. It is a decentralized instrument of power. The cognitive architecture operates as a localized Governance Engine, enforcing corporate doctrine at the very edge of the network. This is the foundational prerequisite for true global expansion.

As the enterprise scales across volatile international borders, adapting to the chaotic, contradictory laws of localized agricultural ministries is an exercise in corporate exhaustion. The sovereign commander refuses to adapt. Instead, we deploy the Governance Engine. Because our cognitive architecture enforces operational thresholds that are vastly more stringent than any existing federal or international law – guaranteeing zero chemical drift, absolute privacy redaction, and acoustic invisibility – we render local regulations irrelevant. We operate in a state of permanent over-compliance.

This is not a defensive posture; it is the aggressive weaponization of technological superiority. By superseding local laws, the enterprise dictates global rules of engagement. We do not integrate into existing frameworks. We force international bodies to recognize our cognitive architecture as the indisputable baseline for agronomic deployment. The technology has matured into an institution.

CHAPTER 42
STRATEGIC INSTITUTIONALISM

42.1 The Treaty Architecture

The corporate world operates on contracts; sovereign states operate on treaties. A contract is a transactional agreement to exchange goods or services, vulnerable to market fluctuations and breach-of-contract litigation. The AgTech enterprise that relies on standard service contracts with regional farms remains a fragile, replaceable vendor. The sovereign commander abolishes the transactional contract and institutes the Treaty Architecture.

When the enterprise engages with global agribusiness conglomerates, national agricultural ministries, or tier-one supply chain distributors, we do not negotiate pricing for drone spraying services. We negotiate infrastructural alignment. We offer absolute, mathematically verified compliance with environmental mandates (SDG 12 and 15), tokenized carbon sequestration liquidity, and real-time phytosanitary defense via our encrypted Black Box matrix. In exchange, the enterprise demands total assimilation.

The partner must integrate their legacy supply chain logistics directly into our proprietary edge-native cognitive network. They must adopt our Zero-Trust cybersecurity protocols and mandate our Variable Rate Application (VRA) algorithms across all their holdings. This is a treaty of technological annexation. The partner receives the biological and financial protection of the swarm, but the enterprise captures their data sovereignty. By structuring these alliances as infrastructural treaties rather than service contracts, we lock the partner into our ecosystem permanently. To break the treaty is not merely to breach a contract; it is to unplug their operations from the biological and financial grid that sustains them.

42.2 Macroeconomic Immunity

The ultimate objective of Strategic Institutionalism is not to beat the market, but to become immune to it. Legacy agricultural businesses are hyper-sensitive to the macroeconomic cycle: when fertilizer prices spike due to geopolitical conflict, or when credit markets tighten, traditional farmers and AgTech startups face immediate margin compression and insolvency. They are tethered to the volatility of the physical supply chain.

The sovereign enterprise engineers Macroeconomic Immunity by decoupling its revenue generation from the raw cost of physical inputs. Because the hexagonal aerospace matrix slashes chemical usage through localized VRA precision, the enterprise is highly insulated against spikes in agrochemical commodity prices. Furthermore, because we have successfully executed Carbon Sequestration Arbitrage (Chapter 39), the enterprise possesses a secondary, uncorrelated stream of capital derived from global carbon markets.

We do not just absorb inflation; we leverage it. When external economic shocks neutralize the profit margins of legacy competitors, forcing them to ground their terrestrial tractors and lay off personnel, the sovereign enterprise utilizes its financial homeostasis to accelerate market capture. We deploy our capital to execute aggressive Mergers and Acquisitions (M&A), buying out distressed regional operators and instantly purging their obsolete hardware to install our proprietary mesh network. Macroeconomic turbulence is not a threat to the institution; it is the exact mechanism by which we rapidly expand our sovereign territory.

42.3 Institutional Intelligence Network

Corporate market research is a reactive autopsy of public data. Legacy agribusinesses and financial analysts wait for quarterly reports from the USDA or international trade commissions to gauge global food supplies and commodity pricing. The sovereign AgTech commander does not wait for public data; they intercept the future. To operate as a true institution, the enterprise must deploy its own sovereign intelligence agency. We construct the Institutional Intelligence Network.

Because our autonomous hexagonal swarms are deployed across diverse international territories under our Treaty Architecture, we do not merely process localized farm data. The edge-native cognitive matrices aggregate hyperspectral and phenotypic telemetry on a hemispheric scale. Long before a national government realizes that a micro-drought or a novel phytosanitary threat is decimating a region's yield, our Black Box architecture has already quantified the exact metabolic deficit of the biological asset.

We possess absolute asymmetric intelligence. By knowing the precise, mathematically verified state of the global food supply before the market does, the enterprise dictates the macroeconomic reality. We leverage this geospatial telemetry to execute aggressive trades in commodity futures and agricultural derivatives. The enterprise effectively operates a private, decentralized intelligence apparatus that transforms global biological data into a weapon of financial foresight, ensuring our capital routing is always months ahead of state-sponsored economic forecasts.

42.4 The Geopolitical Leverage

The ultimate phase of Strategic Institutionalism is the acquisition of absolute leverage over the state itself. When an enterprise provides a luxury good or a standard software service, it is subject to the whims of political regulation and taxation. However, when an enterprise controls the algorithmic autarky that guarantees a nation's food security and mathematically enforces its compliance with global environmental mandates (SDG 12 and 15), the power dynamic is permanently inverted.

The sovereign enterprise engineers systemic dependency. By locking regional cooperatives and national agricultural grids into our encrypted mesh networks and Zero-Trust architecture, we become the central nervous system of their biological survival. If a hostile regulatory body or a foreign government attempts to sanction the enterprise or seize our intellectual property (such as our proprietary Variable Rate Application algorithms or aerospace polymers), we do not engage in standard legal defense. We possess the geopolitical kill switch.

To sever ties with the enterprise is to instantly plunge the region back into thermodynamic inefficiency, chemical cartel dependency, and inevitable crop failure. The state realizes that regulating us out of existence means starving their own biological sectors and failing their international climate treaties. We do not lobby for our right to exist; we establish an architecture of leverage so profound that governments are forced to protect our corporate sovereignty as a matter of their own national security. The institution is absolute.

CHAPTER 43

RED TEAMING & WARGAMING

43.1 The Adversarial Algorithm

A sovereign institution that believes its own corporate rhetoric is already in a state of terminal decline. Legacy agribusinesses and conventional tech startups design their operational models based on optimistic projections – assuming stable weather patterns, uninterrupted supply chains, and static regulatory environments. This is a fatal cognitive bias that breeds systemic fragility. The sovereign AgTech commander operates under a radically different paradigm: the assumption that the world is inherently hostile, deeply volatile, and actively attempting to dismantle the enterprise. To achieve true operational antifragility, the organization must proactively engineer its own destruction before the market or a hostile actor can. We do not hire external consultants to audit our security; we deploy the Adversarial Algorithm.

Red Teaming, within the strict context of The Physics of Capital™, transcends traditional IT penetration testing. It is the institutionalization of an internal, autonomous adversarial force whose sole mandate is to obliterate the enterprise's Operational Homeostasis. We allocate a dedicated, isolated partition of our edge-native supercomputing matrix to act as the apex predator of our own ecosystem. This algorithmic Red Team continuously simulates catastrophic tactics, techniques, and procedures (TTPs) against our proprietary mesh network, our Variable Rate Application (VRA) hardware, and our tokenized carbon ledgers.

The Adversarial Algorithm does not sleep. It launches simulated zero-day exploits against the Black Box, attempting to poison the biological yield predictions by injecting false hyperspectral data into the neural network. It simulates mass GNSS spoofing attacks, feeding fraudulent geospatial coordinates to hijack the hexagonal aerospace swarms and force a kinetic collision. It extends beyond software, executing simulated social engineering attacks on our logistics operators and fabricating sudden supply chain embargoes on the aerospace-grade polymers required for our damping mounts.

By forcing the organization to operate under continuous, simulated siege, the enterprise completely eradicates operational complacency. The swarm learns to navigate, spray, and protect the biological asset not in sterile, ideal laboratory conditions, but in the crucible of absolute systemic chaos. We bleed in the simulation so the harvest never bleeds in the furrow.

43.2 Wargaming the Macroeconomic Collapse

Wargaming is the highest form of strategic foresight. While traditional parcel farmers pray for rain and legacy CEOs lobby for temporary government subsidies during a crisis, the sovereign commander has already played out the exact scenario a million times in the simulated ether. We utilize the vast computational power of the autonomous matrix to wargame the total, simultaneous collapse of both the biological and macroeconomic environments. This is the exact threshold where precision agriculture evolves into predictive warfare.

We construct hyper-realistic, multi-variable simulations where the enterprise is assaulted on all operational fronts. We run Scenario Alpha: A sudden geopolitical conflict in the Pacific Rim severs 80% of the global lithium and rare-earth mineral supply, instantly paralyzing the production of our high-discharge battery architecture, while simultaneously, a novel, aggressive phytosanitary plague strikes the high-value specialty assets of our most critical partners. We follow it with Scenario Beta: A hostile nation-state executes a coordinated cyber-attack on the global commodities market, collapsing the price of our tokenized carbon derivatives, while domestic legacy tractor cartels successfully lobby the EPA to unexpectedly revoke all BVLOS (Beyond Visual Line-of-Sight) permits for autonomous swarms.

In these brutal wargames, the human command structure and the edge-native AI must collaborate in real-time to find the mathematical path to survival. The enterprise maps out contingency logistics, tests the rapid deployment of decentralized, localized energy grids, and practices algorithmic amputation – the ruthless, instantaneous sacrifice of compromised drone units to save the broader integrity of the mesh network. We intentionally stress-test the psychological resilience of our Aviation Directors, forcing them to execute complex capital routing decisions under extreme, simulated duress. These wargames prove that our Macroeconomic Immunity is not a theoretical concept; it is an engineered reality. By subjecting our biological hedging strategies and our supply chain autarky to the absolute worst-case parameters, we guarantee that when the geopolitical reality fractures, the sovereign enterprise does not panic. It simply executes pre-calculated protocols.

43.3 DIN SPEC 91489 and the Anatomy of Resilience

To prevent wargaming from devolving into chaotic, unstructured brainstorming or theatrical corporate exercises, the enterprise subjects all adversarial simulations to the rigid, uncompromising framework of DIN SPEC 91489. We do not view this standard as a mere European guideline or a bureaucratic checklist; we leverage it as the supreme corporate law of resilience. DIN SPEC 91489 demands meticulously structured scenarios, iterative refinement processes, the integration of diverse analytical perspectives, and absolute, forensic documentation. The sovereign enterprise translates these theoretical principles into cryptographic reality.

Every wargame and Red Team assault is treated with the gravity of a live-fire military exercise, strictly adhering to the preparation, execution, and evaluation phases dictated by the standard. The integration of diverse perspectives, a core pillar of DIN SPEC 91489, is executed as a tactical necessity. We force our aerospace engineers, our biochemists, our legal compliance officers, and our quantitative financial analysts into the exact same simulated combat zone. The simulation requires them to cross-pollinate their expertise to survive. We demonstrate that our elastomeric-damped propulsion systems and edge-native cognitive algorithms achieve a level of operational precision that legacy equipment simply cannot replicate. Furthermore, the DIN SPEC 91489 mandate for continuous learning and meticulous evaluation is hardcoded directly into the enterprise's blockchain. The outcomes of every failed strategy, every simulated crashed drone, every chemical spill, and every biological loss are mathematically quantified and cryptographically hashed into the Black Box's permanent memory. This creates a ruthless, Darwinian machine-learning loop. The cognitive architecture of the swarm analyzes these thousands of historical, simulated failures and automatically rewrites its own defense protocols to close the vulnerabilities.

Resilience, under this architecture, is stripped of human emotion. It is no longer about morale or determination; it is a compounding, algorithmic asset. By institutionalizing the DIN SPEC 91489 framework into the very silicon of our operations, the enterprise ensures that it never makes the same mistake twice. The wargame ceases to be an event and becomes a continuous state of evolution, forging a corporate structure that is mathematically impervious to strategic surprise and operational shock.

CHAPTER 44
THE VALLEY OF DEATH

44.1 The Mathematics of Mortality

In the sector of deep technology and physical industrial scaling, the "Valley of Death" is not a metaphor. It is a mass grave of brilliant engineering. Venture capitalists revere software because code scales infinitely at zero marginal cost. Physical infrastructure does not.

The data is violently absolute. Over 80% of hardware and deep-tech ventures die during the transition from prototype to mass production. According to supply chain post-mortems, 75% of venture-backed manufacturing startups face critical insolvency within 24 months of attempting international scale. Why? Because they operate under the fatal delusion of globalized stability. They naively assume that international markets for raw precursors, structural polymers, and localized transport will remain eternally fluid and frictionless.

When a geopolitical crisis erupts, shipping lanes freeze, and raw material costs spike by 300%, software companies adjust their server loads. Hardware and physical infrastructure companies simply bleed out. They are instantly starved of the physical atoms required to sustain their operational nodes. The sovereign enterprise refuses to externalize its survival to a fragile, civilian supply chain. To cross the Valley of Death, we engineer an architecture that mathematically immunizes the operation against global mortality rates.

44.2 Absolute Supply Chain Autarky

You cannot build a sovereign empire while leasing your survival from global spot markets. It is financial suicide to possess the most advanced edge-native cognitive algorithms if your deployment platforms are paralyzed by a shortage of fundamental chemical precursors or energy storage matrices.

The enterprise executes Absolute Supply Chain Autarky. This requires a radical transition: moving from commercial dependency to strategic decoupling. We utilize our Treaty Architecture to secure supply nodes directly at the source, bypassing speculative brokers and international wholesalers. By bringing the critical choke points of raw material acquisition under centralized cryptographic control, the organization insulates itself from external shockwaves. Autarky is not an isolationist philosophy. It is the ultimate risk mitigation protocol. When borders close and legacy competitors cannot source parts to repair their obsolete physical assets, the sovereign enterprise continues to manufacture, deploy, and dominate.

44.3 The Digital Twin Panopticon

It is impossible to defend a supply chain you cannot see. Traditional operations directors rely on delayed spreadsheets, fragmented tracking numbers, and purely reactive crisis management. This analog approach guarantees operational blindness. The sovereign commander eliminates this vulnerability by engineering a Digital Twin Panopticon.

We construct a flawless, virtual replica of the entire global supply network. Every ounce of composite material, every energy cell, and every unit of biological payload is registered immutably on the distributed ledger and projected into the simulated environment. This enables absolute predictive logistics.

If the Institutional Intelligence Network detects a macroeconomic disturbance or a bottleneck at a critical transit point, the system does not wait for the impact to hit the assembly lines. The Digital Twin instantly simulates the scenario, calculates the exact threshold of localized contingency buffers, and autonomously reconfigures the routing of materials to pre-secured secondary geographic nodes. We front-run the crisis. By the time the global market reacts and input prices skyrocket, our structure has already adapted, securing its assets and freezing operational costs. The supply chain ceases to be a sequence of transportation. It becomes a pre-calculated mathematical orchestration.

44.4 Closed-Loop Asset Regeneration

Conventional corporate literature preaches the “circular economy” as an exercise in public relations or environmental guilt. In the physics of capital, circularity is leveraged under the concept of Closed-Loop Asset Regeneration. We do not recycle to appease activists. We regenerate because allowing strategic materials to escape into external landfills constitutes a catastrophic capital leak and a direct violation of autarky.

The operational cycle of any high-capacity autonomous infrastructure involves severe thermodynamic stress. Structural components and energy storage architectures undergo continuous entropic fatigue and progressive performance degradation. In a linear supply chain, damaged components are discarded and replaced via external purchases, continuously draining cash flow. The sovereign enterprise disables this cycle through a closed system where obsolete or degraded hardware is cannibalized and processed internally.

Through predictive data analytics, components approaching their tolerance limits are retired before catastrophic failure. Once recovered, the base elements are refined and reintroduced into the production line. By fusing cryptographic traceability with autonomous manufacturing, the organization accounts for every atom within its ecosystem. The capital invested in raw materials never leaves the structure; it remains trapped in a perpetual loop of value. This level of self-sufficiency severs dependency on external markets, guaranteeing long-term operational viability.

44.5 Asymmetric Crisis Acceleration

The Valley of Death expands exponentially during periods of macroeconomic contraction. Hyperinflation of physical inputs and tightening regulatory policies fracture logistics models based on "just-in-time" principles. Faced with such scenarios, traditional corporations halt expansion, slash budgets, and enter a state of defensive paralysis. The sovereign enterprise does not defend itself against a crisis. It uses the crisis to execute Asymmetric Acceleration.

A crisis is a market-clearing mechanism. Because the organization operates under a regime of operational homeostasis and supply autarky, external shocks fail to penetrate the hull. While competitors hemorrhage capital attempting to sustain inefficient operations, the enterprise's closed-loop continues to provide stable resources at a fixed, pre-calculated cost.

Macroeconomic collapse transforms into the perfect opportunity for monopolistic expansion. The enterprise deploys its capital reserves to absorb the biological assets and regional contracts of insolvent operators. We dismantle their obsolete infrastructure, liquidate their liabilities, and annex their territories directly into our digital network. The chaos that optimizes the traditional industry is the exact fuel the sovereign commander uses to accelerate the business's escape velocity, consolidating a dominance that becomes mathematically irreversible.

CHAPTER 45
ASYMMETRIC EXPANSION

45.1 THE NICHE AS A GEOPOLITICAL BEACHHEAD

The conventional doctrine for scaling a business is linear, predictable, and fundamentally weak: raise capital, launch a broad marketing campaign, and attempt to capture a fractional percentage of a massive, established market. In the sector of autonomous technology and biological management, this symmetric approach guarantees a war of attrition against legacy heavy-machinery cartels equipped with infinite lobbying power and entrenched distribution networks. The sovereign commander does not engage in symmetric warfare. We do not assault the center of the market. We execute Asymmetric Expansion.

Asymmetric expansion dictates that the enterprise must identify an operational niche so highly specialized, so mathematically complex, and so environmentally critical that legacy competitors lack the thermodynamic and algorithmic architecture to even participate. We do not target generic, flat-acreage crop monitoring. That is a commoditized service for amateur operators. Our beachhead is the zero-tolerance, ultra-low volume application of inputs over high-value, topographically hostile biological assets.

By dominating this extreme margin, the enterprise establishes absolute technological supremacy. Legacy heavy machinery cannot navigate the terrain without optimizing the subsoil macroporosity, and standard civilian tech lacks the payload stability, the advanced structural matrices, and the edge-native cognitive algorithms required to prevent toxic drift. We isolate a market segment where our technology is not just an alternative; it is the only mathematically viable option.

The niche is not a limitation. It is a fortified beachhead from which the enterprise launches its total market invasion.

45.2 The Algorithmica Franchise

Once the beachhead is secured, rapid global scaling becomes a problem of capital expenditure (CAPEX) and human logistics. Traditional agricultural service companies attempt to scale by purchasing infinite fleets of hardware and hiring thousands of local operators. This model burns capital, dilutes operational quality, and shatters homeostasis. The sovereign enterprise scales asymmetrically by transitioning from a hardware operator to a software and logistics franchisor. We execute the Algorithmic Franchise.

We do not export hardware; we export the ecosystem. The enterprise partners with localized cooperatives, regional aviation directors, and global logistics hubs. The local partner absorbs the CAPEX of the physical deployment platforms and the labor cost of the certified operators. In return, the enterprise licenses the Black Box – the encrypted artificial intelligence, the proprietary fluid dynamics algorithms, and the blockchain-verified regulatory compliance moat.

This is the ultimate asymmetric scaling mechanism. The enterprise expands its global footprint and aggregates petabytes of telemetry across multiple continents without bleeding its own capital. The local operator acts as the physical actuator, while the sovereign commander retains absolute control over the cognitive architecture and the data derivatives. We capture the high-margin, recurring software and data revenue, while the localized franchisee navigates the friction of local labor laws. The enterprise scales its intelligence infinitely, unburdened by the gravity of physical logistics.

45.3 The Human Capital Moat

In an industry pivoting toward autonomous networks and edge-native supercomputing, the value of the human operator is not diminished;

it is radically redefined. As the enterprise executes the Algorithmic Franchise, the primary operational risk shifts to the localized pilot. A mathematically perfect deployment platform cannot compensate for an operator who lacks the tactical discipline to manage environmental variables or the regulatory knowledge to navigate airspace compliance. Therefore, the enterprise leverages education.

We do not rely on generic, state-sponsored aviation licenses to qualify our operators. We establish a proprietary, draconian Training and Certification Academy. To operate within our licensed ecosystem, field technicians must pass our internal cryptographic, regulatory, and biological stress tests. They are trained not as mere operators, but as Biological Systems Managers.

This proprietary certification process serves a dual asymmetric purpose. First, it ensures that every franchisee operates the technology at the exact thermodynamic tolerances dictated by the Black Box, protecting the brand's liability and preserving the integrity of the ecosystem. Second, it creates a massive Human Capital Moat. By establishing our training protocols as the absolute gold standard of precision agriculture, we force the global labor market to adapt to our operational language. When technical universities begin teaching our proprietary certification standards just to ensure their graduates are employable, the enterprise has successfully monopolized not just the technology, but the very minds of the workforce.

45.4 Systemic Dependency

A standard software-as-a-service (SaaS) model is a fragile civilian construct. Tech startups boast about their recurring revenue, yet their clients can cancel a subscription with a single click and migrate to a competitor. The sovereign enterprise does not build software subscriptions. We engineer absolute systemic dependency. When a localized cooperative or a global agribusiness enters our Algorithmic Franchise, we execute the Biological Lock-In.

The moment a partner integrates their acreage into our edge-native cognitive network, their entire operational reality changes. Our Black Box architecture optimizes their application algorithms, elevating their yield while simultaneously generating cryptographic proof of their ecological compliance. This data is instantly tokenized, creating a secondary stream of high-velocity liquidity for the partner.

If that franchisee decides to leave our ecosystem and return to legacy machinery or amateur operators, they do not just lose a software interface. They instantly lose their tokenized carbon liquidity, their compliance with environmental mandates evaporates, and their biological asset is once again exposed to toxic drift and macroeconomic volatility. The switching cost is no longer a matter of software migration; it is financial and biological suicide. We do not need to lock our partners into restrictive legal contracts. The sheer thermodynamic and economic superiority of the ecosystem traps them voluntarily.

45.5 Predatory Annexation

Expansion is not merely the acquisition of new territory; it is the active consumption of the obsolete. As the macroeconomic environment tightens – driven by inflation, shifting climate patterns, and draconian environmental regulations – traditional service providers and legacy hardware operators inevitably face insolvency. They possess valuable regional contracts and access to prime biological assets, but they are bleeding capital to maintain their heavy, inefficient machinery. The sovereign commander watches this decay and executes Predatory Annexation.

We do not acquire distressed competitors to absorb their operations; we acquire them to liquidate their past. When the enterprise buys out a failing regional operator, we do not integrate their outdated hardware into our fleet. That hardware is a liability. We execute an immediate, ruthless asset purge. The legacy machinery is scrapped, sold, or recycled into our closed-loop asset regeneration system.

We strip the acquired company down to its core value: its land access and its client contracts. We then instantly deploy our proprietary autonomous networks over the newly acquired territory. By stripping away the massive operational expenditure (OPEX) of the legacy machinery and replacing it with our algorithmic autarky, an acquisition that was bankrupt on Monday becomes a highly profitable, self-sustaining node in our empire by Friday.

45.6 The ESG Liquidity Vacuum

The final vector of asymmetric expansion occurs far above the soil; it takes place in the boardrooms of institutional finance. To scale infinitely, an enterprise must not only secure its own capital but actively starve its competitors of theirs. Trillions of dollars are locked within institutional Environmental, Social, and Governance (ESG) funds, mandated to invest strictly in verifiable, sustainable technologies.

Legacy heavy-machinery cartels and conventional monopolies cannot access this capital because their operations are inherently destructive to the subsoil and the atmosphere. The sovereign enterprise captures this liquidity entirely. Because our cognitive architecture generates unforgeable cryptographic hashes proving absolute ecological silence, zero-drift precision, and carbon sequestration, we become the only mathematically viable investment for institutional green capital.

We execute Capital Starvation. By monopolizing the flow of ESG venture capital and institutional funding, we deprive legacy competitors of the financial oxygen they need to conduct research and development. They cannot afford to build edge-native AI or aerospace-grade structures because the global financial markets refuse to fund their obsolete, carbon-heavy business models. We do not just out-engineer the competition in the field; we bankrupt them on Wall Street. Asymmetric expansion guarantees that as the sovereign enterprise accelerates toward escape velocity, the rest of the industry is left choking on the fumes of a defunded, dying paradigm.

CHAPTER 46
THE MINERAL FRONTIER

46.1 The Geopolitics of the Crust

A fatal delusion plagues the modern AgriTech industry: the belief that software eats the world. The sovereign commander understands a much darker, older truth: geopolitics eats software. You can engineer the most advanced edge-native artificial intelligence, write flawless Variable Rate Application (VRA) algorithms, and encrypt your mesh networks behind unassailable Zero-Trust architecture. None of it matters if the physical atoms required to build your machines are controlled by a hostile foreign power.

The ultimate vulnerability of any aerospace enterprise is the Earth itself. The heavy-lift hexagonal swarms that execute our biological interventions are physical manifestations of rare-earth chemistry. The proprietary high-torque propulsion architecture utilizes rare-earth-based magnetic induction arrays to generate the required mechanical output. The energy matrices require lithium, cobalt, and nickel. Currently, global supply chains are dangerously asymmetric; a single geopolitical bloc controls over 85% of rare-earth refinement and 70% of lithium-ion battery production. If a trade war erupts, or a naval blockade chokes the critical shipping straits, civilian startups will be instantly starved of hardware.

The sovereign enterprise refuses to exist at the mercy of global commodity cartels. We execute a strategy of Deep Vertical Integration, bypassing the volatile spot markets to anchor our supply chain directly to the bedrock. We do not just build the drone; we secure the crust from which it is forged. By establishing off-take agreements and strategic treaties with localized, allied mining cooperatives, we lock in our raw material OPEX years in advance. We integrate the Earth itself into our corporate ledger, mathematically ensuring that no international tariff or border closure can ever ground the sovereign fleet.

46.2 Remote Sensing as Commodity Warfare

To secure the mineral frontier, the enterprise does not rely on external geological surveys. We leverage the very technology we manufacture. The same autonomous hexagonal swarms that read the phenotypic telemetry of a high-value specialty asset are instantly repurposed for high-altitude geological reconnaissance. This is the synthesis of precision agriculture and commodity warfare.

By equipping the aerospace matrix with Short-Wave Infrared (SWIR) sensors, advanced LiDAR, and multispectral imaging, the swarm executes autonomous mineral exploration across vast, uncharted topographies. The drones read the electromagnetic reflectance signatures of the earth, identifying the specific spectral anomalies that indicate subsurface deposits of lithium brines, copper, and essential agricultural silicates.

We generate proprietary, high-resolution geological maps that legacy mining corporations spend tens of millions of dollars to acquire. The enterprise then uses this asymmetric data to execute brutal cost-benefit analyses. We approach raw-material suppliers not as desperate buyers, but as intelligence brokers. We trade our localized geological telemetry for guaranteed, discounted access to their physical yields. By transforming our agricultural drones into geological oracles, the enterprise ensures a perpetual, self-sustaining loop: the drone maps the minerals, the minerals build the battery, the battery powers the drone. The supply chain becomes an unbreakable circle of thermodynamic independence.

46.3 The Lithium Beachhead and Energy Density

The deployment of heavy-lift aerospace platforms in agriculture is fundamentally an equation of energy density. Hovering a drone with a 40-liter liquid payload requires massive, continuous kinetic output. Traditional operators utilize generic lithium-polymer (LiPo) packs, which degrade rapidly under extreme thermal stress, forcing the operator to constantly repurchase batteries at highly inflated retail prices. This degrades the Total Cost of Ownership (TCO) and optimizes financial homeostasis.

The sovereign commander establishes a Lithium Beachhead. Because we have secured our raw chemical precursors through our geological treaties, the enterprise is not subject to the hyperinflation of the battery market. We aggressively engineer our own power cells, pushing the boundaries of solid-state architecture to achieve energy densities exceeding 350 Wh/kg.

By commanding the proprietary chemical architecture of our energy storage systems, we dictate the exact thermal degradation trajectory of our autonomous infrastructure. When global lithium prices spike by 300% due to supply chain fractures, legacy tractor companies and amateur drone pilots are paralyzed by the cost of operation. The sovereign enterprise, insulated by its strategic reserves and proprietary energy matrices, feels nothing. We continue to deploy our swarms at a fixed, mathematically verified OPEX. Securing the mineral frontier ensures that while the rest of the market drowns in the volatility of commodity super cycles, the sovereign enterprise accelerates its global expansion, fueled by the very earth it dominates.

46.4 The Cannibalization of Capital

In the endgame of resource acquisition, extraction is not limited to virgin soil. The sovereign enterprise recognizes that the corpses of bankrupt legacy agribusinesses provide a massive, untapped vein of elemental capital. Traditional operators view obsolete hardware as depreciated liabilities – junk to be sold for scrap or abandoned in warehouses. The AgTech commander views the graveyard of the competition as an anthropogenic mine. We execute Urban Mining through the ruthless Cannibalization of Capital.

When Predatory M&A (Chapter 45) absorbs a failing competitor, we acquire fleets of inferior quadcopters and heavy terrestrial tractors. We do not fly them. We melt them down. The enterprise deploys automated disassembly nodes that strip the neodymium from their inefficient motors, recover the copper from their winding coils, and chemically leach the lithium and cobalt from their bloated, degraded battery packs.

This process fundamentally alters the Total Cost of Ownership (TCO) equation. By extracting rare-earth elements from the failures of the market, we bypass the immense capital expenditure (CAPEX) required for primary geological mining. We recycle the enemy's hardware to feed our own closed-loop asset regeneration system. The competitor's bankruptcy directly subsidizes the manufacturing of our next-generation hexagonal swarms. We do not just defeat the competition; we physically consume their atoms to build our empire.

46.5 The Macromolecular Stronghold

While rare-earth elements dictate the kinetic output of the aerospace matrix, the structural survival of the drone relies entirely on its macromolecular skeleton. The proprietary elastomeric damping mounts and aerospace-grade polymers are what allow the hexagonal airframe to absorb the brutal mechanical fatigue of continuous, high-capacity liquid spraying. If a global petrochemical bottleneck starves the enterprise of these specific precursor resins, the fleet shatters under its own torque.

The sovereign commander anticipates this chemical vulnerability and establishes a Macromolecular Stronghold. We refuse to be held hostage by international polymer distributors who prioritize military or automotive contracts over agricultural technology. The enterprise secures direct upstream alliances with chemical refineries, dictating the exact monomer synthesis required for our proprietary components.

By taking command of the chemical formulation at the molecular level, we dictate the exact tensile strength, thermal resistance, and vibrational absorption of the hardware. When international supply chains fracture and the cost of generic industrial plastics skyrockets, competitors are forced to ground their fleets or risk catastrophic mid-air structural failures. The sovereign enterprise operates seamlessly, insulated by a macromolecular shield. We control the chemistry, therefore we control the physical reality of the airspace.

46.6 Elemental Sovereignty

The completion of the Mineral Frontier marks the absolute decoupling of the enterprise from the chaos of the global physical economy. An organization that relies on the spot market for its lithium, neodymium, and aerospace polymers is merely a tenant in its own industry, paying rent to global commodity cartels. By mapping the earth with our own autonomous sensors, extracting raw materials through strategic treaties, and cannibalizing the hardware of dead competitors, we eradicate the concept of the external supplier.

This is Elemental Sovereignty. Every atom within the corporate ecosystem is mapped, owned, cryptographically tracked, and infinitely recycled. The enterprise creates a self-sustaining thermodynamic loop that defies the entropy of the traditional macroeconomic cycle. We have successfully secured the software, the biology, the law, and now, the bedrock. The sovereign enterprise is no longer a participant in the global economy; it is a self-contained, impregnable physical universe.

CHAPTER 47
COMMODITY SUPER-CYCLE

47.1 The Physics of the Super Cycle

In the vocabulary of legacy economics, a Commodity Super Cycle is defined as a prolonged, macroeconomic period where the demand for raw materials significantly outstrips the global supply, driving prices to historic, sustained highs. Traditional farmers, regional supply chain managers, and standard AgTech startups view these cycles with passive anxiety. They attempt to optimize their basic agronomic practices – crop rotation, soil health, and generic pest management – hoping to survive the crushing inflation of inputs while praying the spot market for their yield remains favorable. This is a strategy of profound weakness.

The sovereign commander does not survive the super cycle; they engineer their enterprise to consume it. The Physics of Capital™ dictates that a super cycle is not an unpredictable act of economic weather; it is a thermodynamic transfer of wealth. When global populations surge toward 10 billion and the supply of arable land and clean water collapses under the weight of climate volatility, the demand for both raw agricultural commodities and the rare-earth minerals required to farm them autonomously will skyrocket exponentially.

Legacy competitors are instantly paralyzed during the upswing of a super cycle. Their entire operational model relies on purchasing synthetic chemical inputs, generic lithium batteries, and heavy diesel-powered terrestrial tractors at spot-market prices. When those prices triple, their OPEX (Operational Expenditure) exceeds their revenue, and they default. The sovereign enterprise, however, has already executed Absolute Supply Chain Autarky (Chapter 44) and secured the

Mineral Frontier (Chapter 46). By decoupling our operations from the spot market and controlling our own macromolecular and rare-earth inputs, our OPEX remains locked and mathematically flat. As the super cycle drives the global price of food and agricultural commodities to historic highs, our profit margins do not just grow – they explode asymmetrically. The super cycle becomes the gravity well that pulls the wealth of the dying legacy sector directly into our treasury.

47.2 Soil Health as a Derivate Asset

The provided civilian literature correctly identifies that soil health – specifically organic matter content, microbial activity, and macroporosity – is vital for sustainable yield and resilience against climate change. However, it frames soil health as an ecological duty or a traditional agronomic skill. The sovereign enterprise reframes soil health as a high-velocity financial derivative.

During a commodity super cycle, the physical yield of the crop is only the first layer of monetization. The true asymmetric leverage lies in the ground itself. Because the legacy agricultural sector has spent decades annihilating soil biomes with synthetic chemical inputs and heavy tractor compaction, fertile, carbon-rich soil has become an extremely scarce global asset.

Our autonomous application platforms do not touch the ground. Our zero-drift Variable Rate Application (VRA) algorithms apply ultra-precise, proprietary biological controls (biopesticides) that actively regenerate the subsoil microbiome. As the swarm flies, the edge-native cognitive architecture generates continuous, cryptographic phenotypic telemetry (Chapter 39) that proves the soil's regenerative state.

We do not just harvest the crop; we harvest the data. During the super cycle, as global corporations desperately seek to offset their carbon emissions to comply with draconian environmental regulations, the sovereign enterprise tokenizes this mathematically verified soil health. We transform the microscopic biological activity of the dirt into a tradable financial asset on the global carbon exchange. While the traditional farmer struggles to sell their physical harvest against currency fluctuations, the AgTech commander sells the cryptographic proof of the Earth's respiration, generating massive, uncorrelated liquidity that further subsidizes the expansion of the drone fleet.

47.3 The Climate Arbitrage

Climate change is universally classified by governments, international organizations, and traditional farmers as an unmitigated disaster – an unpredictable force that optimizes crop yields through extreme droughts, floods, and shifting pest dynamics. Consequently, global policy responses focus on defensive adaptation strategies and subsidies for failing farms. The sovereign enterprise views climate volatility through a radically different lens: it is the ultimate engine for Climate Arbitrage.

Arbitrage is the exploitation of price differences across disparate markets. In the context of the commodity super cycle, climate change creates localized zones of total agricultural failure alongside zones of temporary surplus. Because the enterprise has deployed its Franchise of Algorithms (Chapter 45) globally, our Institutional Intelligence Network (Chapter 42) aggregates real-time environmental data across hemispheres.

When our predictive yield oracles detect that a shifting climate pattern will devastate the crop output of a specific geographic region, the enterprise does not merely adapt its local farming practices. We execute financial warfare. Armed with this asymmetric intelligence, the enterprise aggressively shorts the commodity futures of the failing region while simultaneously taking long positions in the regions where our swarms mathematically guarantee a high yield.

Furthermore, as climate volatility disrupts traditional pest cycles, the enterprise leverages its proprietary advancements in biological pest control. We deploy our precision swarms to deliver targeted microbial and bio-based interventions exactly where the climate shift has rendered traditional chemical pesticides useless. We monetize the ecological chaos. By maintaining an agile, autonomous, and cryptographically shielded infrastructure, the sovereign commander does not suffer the effects of climate change; they calculate the exact trajectory of the collapse and position the enterprise to absorb the capital displaced by the disaster.

47.4 Currency Weaponization and Forex Immunity

In the global agricultural theater, the silent superior of the traditional farmer is not drought or pests; it is foreign exchange (Forex) volatility. During a commodity super cycle, inflation rages, and emerging market currencies often collapse against apex currencies like the US Dollar or the Euro. The legacy farm owner is caught in a decisive economic trap: they are forced to import synthetic fertilizers, diesel fuel, and proprietary seeds priced in strong foreign currencies, while selling their physical yield in weak, depreciating local fiat. This currency mismatch annihilates their profit margins before the seeds even touch the soil.

True empathy for the agricultural sector demands that we build an enterprise mathematically immune to this slaughter. The sovereign commander does not beg central banks for favorable exchange rates; they engineer Forex Immunity.

Because we have executed Absolute Supply Chain Autarky (Chapter 44) and secured our raw materials at the bedrock (Chapter 46), our reliance on imported, dollar-denominated agrochemicals and hardware components drops asymptotically toward zero. But the strategy goes further: we leverage the currency disparity. The enterprise operates localized autonomous nodes in emerging markets, paying the fractional Operational Expenditure (OPEX) in depreciated local currency. Simultaneously, the unforgeable proof of carbon sequestration and biological yield generated by our edge-native cognitive network is tokenized and sold on international exchanges in apex currencies.

We generate high-value, hard-currency revenue streams from the exact same acreage that costs us pennies in local fiat to operate. The enterprise effectively transforms the volatility of the global currency markets into an aggressive arbitrage engine, extracting maximum capital from geopolitical instability.

47.5 The Tariff Shield and Diplomatic Passports

As a commodity super cycle accelerates, national governments panic. Driven by the fear of food shortages and inflation, states rapidly alter trade policies, erecting massive tariffs, enforcing export bans, and engaging in protectionist trade wars. The traditional agricultural exporter is battered by these sudden bureaucratic walls, their products trapped at borders or rendered uncompetitive by overnight taxes.

The AgTech commander understands that in a hyper-regulated global economy, the ultimate trade commodity is not the crop itself, but compliance. We deploy the Tariff Shield.

When international trade ministries lock down borders, they leave deliberate loopholes for products and technologies that help them meet their binding international climate commitments, such as the Paris Agreement and the UN Sustainable Development Goals (SDGs). Because our aerospace matrix applies bio-based inputs with absolute zero-drift precision, preserving soil macroporosity and eliminating watershed toxicity, our yield is cryptographically certified as apex-tier sustainable.

We do not export mere "agricultural goods." We export algorithmic, verifiable ecological salvation. When our partner networks ship their yield across borders, it is accompanied by the Black Box's immutable blockchain ledger, proving its carbon-negative footprint. Governments do not tax this; they subsidize it. They wave it through customs because importing our cryptographically verified yield allows the host nation to artificially bolster their own domestic sustainability metrics. The cognitive architecture of the swarm becomes a diplomatic passport, rendering international trade barriers completely frictionless for the sovereign enterprise while locking out legacy competitors.

47.6 The Centrifuge of Wealth

The culmination of a commodity super cycle separates the architects of capital from the victims of circumstance. Legacy agribusinesses view the super cycle as a violent storm to be endured. They take on massive debt to cover the hyperinflation of inputs, hoping that the high price of commodities will bail them out at the end of the harvest. But a super cycle is not a storm; it is a macroeconomic centrifuge.

The immense rotational G-force of the cycle spins out the inefficient, the chemical-dependent, and the technologically obsolete. As traditional operators are crushed by rising input costs, currency collapses, and hostile trade policies, their prime biological assets (the land) and market share are stripped away.

The sovereign enterprise stands perfectly still at the dead center of the centrifuge. Shielded by our Macromolecular Stronghold, powered by the continuous flow of our closed-loop asset regeneration, and armed with our predictive biological oracles, we feel no gravity. As the legacy operators are spun into bankruptcy, the enterprise deploys its capital to execute predatory acquisitions, absorbing their territories and purging their obsolete hardware.

The super cycle is the ultimate mechanism of the Commander's design. We do not just survive the era of scarcity and inflation; we use its violent momentum to finalize our total hegemony over the agricultural and financial landscape.

CHAPTER 48

ADVANCED FINANCIAL ENGINEERING

48.1 The Alchemy of Capital

To the uninitiated, finance is a system of accounting – a method for tracking revenue, managing debt, and projecting quarterly growth. Legacy CEOs and traditional agribusiness directors operate within this paradigm, attempting to balance their ledgers against the brutal realities of the macroeconomic cycle. This is a defensive, civilian posture. The sovereign AgTech commander recognizes that true financial engineering is not accounting; it is alchemy. It is the deliberate, mathematical transmutation of physical reality into asymmetrical capital leverage.

The core doctrine of The Physics of Capital™ states: "Thesystemic elimination of the competitor is not a goal; it is a thermodynamic byproduct of absolute algorithmic autarky and macroeconomic immunity."

In this chapter, we transition from the physical realm of hexagonal swarms and rare-earth minerals into the pure mathematics of financial hegemony. We do not just optimize portfolios; we invent the instruments of our own liquidity. We do not just manage risk; we securitize it and sell it to our competitors. Advanced financial engineering is the ultimate weaponization of our entire physical and digital ecosystem. It is the mechanism by which the enterprise escapes the gravity of traditional agricultural economics and achieves perpetual, self-sustaining financial velocity.

48.2 Algorithmic Capital Routing

Standard corporate finance relies on human committees to allocate capital. Decisions are made quarterly, heavily influenced by behavioral biases, emotional attachments to legacy projects, and reactive panic to market news. In the context of a volatile commodity super cycle, human reaction time is functionally obsolete. The sovereign enterprise replaces the boardroom with Algorithmic Capital Routing.

We integrate advanced Quantitative Portfolio Optimization directly into the edge-native supercomputing matrix that governs the drone fleet. The enterprise does not view its operations as a single farm or a logistics network; it views its global footprint as a massive, high-frequency trading portfolio. Every hectare of land under our Treaty Architecture (Chapter 42), every liter of proprietary biopesticide, and every kilowatt of solid-state battery energy is a tradable asset.

As the hexagonal platforms execute their zero-drift biological interventions, the Black Box continuously streams real-time phenotypic telemetry, soil carbon retention rates, and local micro-climate data back to the central cognitive architecture. Machine learning algorithms, free from cognitive bias, analyze this data against global commodity futures, currency fluctuations (Forex), and supply chain constraints.

If the algorithm detects that the cost of applying biological inputs in Sector A will yield a lower financial return than the spot price of selling those inputs directly to the market, it autonomously halts the deployment, reroutes the capital, and liquidates the inventory. The enterprise executes High-Frequency Trading not on a Wall Street terminal, but on the physical reality of the agricultural supply chain. We optimize the allocation of physical atoms and financial capital at lightning speed, ensuring that Operational Expenditure (OPEX) is always directed toward the absolute highest mathematical yield.

48.3 Securitization of Biological Yield

The true power of structured finance lies in securitization – the process of pooling illiquid, disparate assets and transforming them into standardized, tradable financial instruments. Wall Street revolutionized the global economy by securitizing mortgages. The sovereign AgTech enterprise revolutionizes the global economy by securitizing the Earth's biological yield.

Traditional farming is inherently illiquid. A farmer plants a crop and waits months to monetize the harvest, absorbing the total risk of drought, pests, and price collapses. The sovereign enterprise eliminates this risk by engineering the Biological Collateralized Debt Obligation (B-CDO).

Because our enterprise operates under a regime of Absolute Supply Chain Autarky (Chapter 44) and utilizes Predictive Yield Oracles (Chapter 39), our future harvests are not speculative gambles; they are mathematically guaranteed contracts. We aggregate the projected biological yield across thousands of localized Franchise networks (Chapter 45). We bundle these future yields, backed by the unforgeable cryptographic proof of our drone telemetry, into high-grade financial securities.

These B-CDOs are structured in tranches and sold to institutional investors, hedge funds, and sovereign wealth funds desperately seeking assets uncorrelated to the traditional stock market. The enterprise receives massive, immediate liquidity upfront, long before a single seed has breached the topsoil. We use this capital to fund further predatory M&A and R&D. The risk of the harvest is entirely transferred to the financial markets, while the enterprise retains operational control. We do not farm for profit; we manufacture financial instruments backed by the thermodynamic certainty of our technology.

48.4 Behavioral Alpha and the Monetization of Panic

Financial markets are deeply irrational, governed by the psychological fragility of human traders. In the agricultural commodity sector, this fragility is magnified by the primal fear of starvation, weather anomalies, and geopolitical collapse. Legacy CEOs and institutional investors fall victim to cognitive biases – herd mentality, loss aversion, and overreaction to macroeconomic news. They buy at the peak of climate panic and sell at the bottom of regulatory despair. The sovereign AgTech commander does not participate in this hysteria. We extract Behavioral Alpha from it.

Behavioral Alpha is the excess financial return generated by exploiting the emotional errors of the market. Because the enterprise operates entirely via the cold, calculating logic of the edge-native cognitive architecture, it possesses absolute emotional immunity. But we do not merely observe the panic of our competitors; we actively induce and monetize it.

When our Institutional Intelligence Network (Chapter 42) detects an impending supply chain fracture or a regional yield failure, we do not conceal our own operational stability. We publish the cryptographic proof of our flawless, autonomous biological yields directly to the market. This creates immediate, devastating psychological friction. Institutional capital realizes that legacy farms are failing while the sovereign enterprise is mathematically thriving. Panic accelerates. Investors dump traditional agricultural derivatives and rush to our Biological Collateralized Debt Obligations (B-CDOs). We short the legacy market as it crashes under the weight of its own psychological terror, extracting massive liquidity from the very fear we helped precipitate.

48.5 Decentralized Financial Warfare

Blockchain technology is fundamentally misunderstood by the civilian tech sector. They view it as a decentralized ledger for cryptocurrencies or a transparent database for supply chain tracking. In The Physics of Capital™, the blockchain is not a database. It is an automated executioner.

We deploy Smart Contracts to execute Decentralized Financial Warfare. A smart contract is self-executing code that automatically triggers a financial action when specific mathematical parameters are met. The sovereign enterprise leverages these contracts to eliminate the delays and legal friction of traditional corporate warfare.

Consider the mechanics of our Treaty Architecture (Chapter 42) with regional partners and competitors. We embed smart contracts into the very fabric of our algorithmic autarky. If a legacy competitor attempts to hedge against us, or if a regional operator breaches the thermodynamic tolerances of our ecosystem, we do not file a lawsuit. We do not hire attorneys to engage in multi-year litigation. The smart contract, fed by real-time data from the autonomous matrix, detects the failure. Instantly, autonomously, and without human intervention, the contract executes. It liquidates their collateral, reroutes their tokenized carbon credits directly into our treasury, and cuts off their access to the mesh network.

The enemy defaults before their CEO has even been briefed on the breach. By fusing high-frequency financial execution with edge-native biological telemetry, the enterprise enforces its hegemony at the speed of light. Law is no longer written on paper; it is compiled in silicon and enforced by mathematics.

48.6 The Thermodynamic Pricing Engine

To completely dominate advanced financial engineering, the enterprise must redefine how value is calculated. Traditional quantitative analysts use models like Black-Scholes or Monte Carlo simulations to price financial derivatives based on historical stock volatility and interest rates. This is insufficient for an empire built on physical atoms and biological assets. The sovereign enterprise engineers the Thermodynamic Pricing Engine.

We fuse financial data science with the actual physics of our drone swarms. Our pricing models do not just calculate the volatility of the agricultural market; they calculate the exact energy density of our solid-state batteries, the aerodynamic drag coefficients of our flight paths, and the cellular response of the canopy to our proprietary biopesticides. We calculate risk in Joules and Watts, not just dollars.

By running Monte Carlo simulations that combine macroeconomic variables (currency fluctuations, rare-earth embargoes) with the thermodynamic reality of our hardware, the enterprise creates pricing models of unprecedented accuracy. When we sell futures contracts or price our B-CDOs, we are not guessing. We have mathematically priced the precise amount of energy required to force the biological asset to survive. This fusion of quantitative finance and hard engineering creates a pricing model so absolute that Wall Street quantitative analysts are forced to trade on our terms, using our metrics. We establish the monetary value of physical reality.

48.7 The Agricultural Dark Pool

Legacy commodity exchanges – the Chicago Mercantile Exchange (CME) or the global spot markets – are archaic, public arenas designed for civilians. They suffer from transparency, regulatory friction, and retail latency. To execute high-frequency trading and extract maximum yield from our Biological Collateralized Debt Obligations (B-CDOs), the enterprise cannot rely on external, state-sponsored financial infrastructure. The sovereign commander bypasses Wall Street entirely by engineering the Agricultural Dark Pool.

A Dark Pool is a private financial exchange where institutional trading operates outside the purview of public markets, preventing market impact until the trade is executed. The enterprise constructs the world's first edge-native biological Dark Pool. Because we hold absolute monopolies over the asset (the biological yield) and the intelligence (the phenotypic telemetry), we act as the ultimate, uncontested market maker.

We route the massive liquidity of global ESG funds, sovereign wealth portfolios, and institutional capital directly into our private exchange. Wall Street quantitative analysts and algorithmic trading firms are forced to plug their API keys into our infrastructure, paying us exorbitant latency fees just for the privilege of trading our tokenized carbon credits and B-CDOs. We do not pay fees to brokers; we own the casino. By executing buy and sell orders internally at the speed of light, we create a Liquidity Singularity. The enterprise absorbs global capital so rapidly and frictionlessly that public agricultural markets are drained of volume, leaving competitors trading in a barren, illiquid wasteland.

48.8 The Doctrine of Thermodynamic

It is here, at the absolute apex of financial and physical integration, that the fundamental law of the enterprise is revealed. Civilian CEOs draft mission statements about "defeating the competition" or "disrupting the industry." They engage in brutal price wars, marketing campaigns, and hostile litigation, bleeding their own capital in a desperate attempt to wound their rivals. This is a profound miscalculation of physics and finance.

The core doctrine of The Physics of Capital™ dictates: The systemic elimination of the competitor is not a goal; it is a thermodynamic byproduct of absolute algorithmic autarky and macroeconomic immunity.

In the realm of physics, energy cannot be created or optimized; it can only be transferred. Capital is the economic equivalent of energy. When the sovereign enterprise secures the Mineral Frontier (Chapter 46), we cut off the enemy's access to physical kinetic energy. When we execute Capital Starvation through our proprietary ESG compliance moat (Chapter 45), we drain their financial energy. When we deploy the Thermodynamic Pricing Engine and the Agricultural Dark Pool, we monopolize the global flow of liquidity.

We do not attack the competitor. We simply engineer a localized vacuum of capital and resources. The competitor perishes not because we struck them, but because the enterprise has absorbed all the thermodynamic fuel in the ecosystem. Their terrestrial tractors rust, their legacy supply chains fracture, and their unhedged capital is incinerated by inflation. Their destruction is a silent, mathematical certainty – the inevitable result of entropy acting upon a system that lacks algorithmic autarky. The enterprise achieves monopoly without firing a single shot, simply by watching the competition freeze to death in the financial void we engineered.

48.9 Algorithmic Fiat

The final frontier of financial engineering transcends the accumulation of wealth. To amass billions of US Dollars or Euros is to remain subjugated to the monetary policy, inflation rates, and debt cycles of national central banks. A true sovereign institution does not stockpile foreign fiat; it mints its own. The AgTech commander establishes the Sovereign Treasury.

Because our proprietary blockchain ledger continuously verifies the carbon sequestration of the soil and the exact caloric and thermodynamic output of the global harvest, the enterprise generates the ultimate hard asset. We elevate our internal tokenized derivatives into a form of Algorithmic Fiat. Unlike national currencies, which are backed by nothing but political promises and military debt, our Algorithmic Fiat is backed by the mathematically verified, life-sustaining biological yield of the Earth itself.

The enterprise begins conducting its macroeconomic transactions entirely off the grid of standard currency. We execute Predatory M&A, acquire raw lithium and aerospace-grade polymers, and compensate our localized franchisees using our own biological tokens. As geopolitical instability shatters trust in national fiat currencies, institutional partners and sovereign states recognize the superior stability of an algorithmic currency pegged to actual, physical survival.

The enterprise ceases to be a corporation. It becomes a decentralized central bank. We control the physical harvest, the logistics, the trading exchange, and the currency itself. Absolute macroeconomic hegemony is achieved.

CHAPTER 49

THE MACRO FORECAST

49.1 The Predictive ERP

The greatest vulnerability of a traditional corporate empire is its own nervous system. Legacy agribusinesses, global logistics firms, and civilian CEOs utilize Enterprise Resource Planning (ERP) software – such as SAP or Oracle – as a digital rearview mirror. They use these massive systems merely to log historical inventory, process payroll, and generate retrospective quarterly reports for their shareholders. This is a catastrophic misallocation of computational power. In a volatile commodity super cycle, looking backwards is a death sentence. The sovereign AgTech commander does not use SAP to record history; we engineer it to dictate the future.

We transform the ERP from an accounting database into the Predictive Macro-Engine of the enterprise. We execute a total, frictionless integration of our edge-native drone cognitive architecture directly into the SAP S/4HANA (or equivalent advanced ERP) environment. Every variable in the physical world is instantly digitized and routed into the corporate brain. When an autonomous application node adjusts its Variable Rate Application (VRA) to compensate for a micro-climate wind shift, that exact nanoliter of saved biopesticide is instantly registered in the ERP's inventory module, adjusting the global Operational Expenditure (OPEX) calculations in real-time.

But the true weaponization of the ERP occurs at the intersection of logistics and the Environmental Protection Agency (EPA). Civilian companies fear the EPA; they spend millions on compliance officers to prepare for brutal, manual audits. The sovereign enterprise neutralizes this friction entirely. We configure our SAP architecture to automatically translate the unforgeable cryptographic telemetry of our zero-drift drone swarms into formatted regulatory data. We do not wait for an EPA audit; we grant the EPA a restricted, read-only API key to our compliance ledger.

The government regulators can log in and see mathematical, blockchain-verified proof of our absolute ecological compliance – zero chemical runoff, localized carbon sequestration, and total adherence to HAZMAT transit laws. By feeding the EPA flawless, automated data, we turn the regulatory body into a passive validator of our supremacy. Meanwhile, the ERP's predictive algorithms scan the macroeconomic horizon, forecasting the exact moment when the EPA will crack down on the toxic runoff of our legacy competitors, allowing us to position our capital to absorb their distressed assets the moment they are penalized.

49.2 The Sales Force as Tactical Operatives

A fundamental error in traditional business scaling is the deployment of a conventional sales force. Legacy companies hire charismatic representatives to push products, negotiate discounts, and beg for regional contracts. This introduces human emotion and margin dilution into a system that should be governed purely by physics and finance. The sovereign enterprise does not have a "sales force." We deploy Tactical Integration Operatives.

Our operatives do not sell drones; they sell macroeconomic survival. They are deployed into the field armed directly with mobile access to our Predictive ERP. When an operative sits down with the director of a massive, failing regional agricultural cooperative, they do not pitch features and benefits. They open their terminal and display the prospect's exact thermodynamic and financial leaks.

Detecting Red Flags and Opportunities: The operative uses the ERP to cross-reference global commodity inflation with the prospect's current reliance on imported, dollar-denominated chemical fertilizers. The red flag is glaring: the prospect's OPEX is rising at 14% annually, while their soil macroporosity is collapsing, putting them directly in the crosshairs of an impending EPA violation.

The operative offers the Biological Lock-In (Chapter 45). They present a real-time, SAP-generated simulation proving that by surrendering their legacy tractors and assimilating into our Franchise of Algorithms, the prospect will instantly achieve compliance, slash their chemical costs to near zero, and begin earning tokenized carbon credits. The operative controls the negotiation not through persuasion, but through inescapable mathematical reality. The sales force enforces the Treaty Architecture, ensuring that every new territory we annex operates strictly under our cryptographic parameters.

49.3 Asymmetric Risk Detection: Real Cases & Protocols

To operate an autonomous global supply chain, the enterprise must flawlessly handle anomalies. The Macro Forecast is only as powerful as its ability to detect operational red flags and execute immediate, automated course corrections. The sovereign commander relies on the ERP to filter the noise of the global market and isolate the true asymmetric risks.

Real Case Protocol: The HAZMAT Supply Chain Fracture

- The Scenario: A localized transport strike paralyzes a critical shipping route in a developing nation. Two tons of proprietary, highly concentrated biopesticides (classified under EPA/ international HAZMAT protocols) are stalled in a port facility, threatening the deployment schedule of three regional franchise swarms.
- The Red Flag: The ERP's supply chain module detects the geofence anomaly – the shipping containers have not moved in 48 hours, and local ambient temperatures are rising, threatening the biological integrity of the payload.

- How to Handle It (The Sovereign Response): A civilian logistics manager would panic, make phone calls, and bleed capital on expedited emergency shipping. Our system removes the human entirely.

    1. The SAP system automatically triggers an autonomous risk protocol.
    2. It assesses the expiration parameters of the stalled biopesticide and calculates the financial penalty of the delayed harvest.
    3. Concurrently, the ERP's Digital Twin Panopticon (Chapter 44) identifies a secondary, localized cache of biopesticides currently held by a neighboring franchise operating at a surplus.
    4. The system automatically executes a smart contract to purchase the surplus from the neighbor, reroutes the deployment drones to pick it up, and legally reclassifies the stalled port inventory for secondary market liquidation once the strike ends.
    5. An automated compliance report is instantly generated and filed with the EPA and local equivalents, documenting the precise chain of custody for the delayed HAZMAT materials, ensuring zero regulatory liability.

The crisis is averted in milliseconds. The biological asset is sprayed on schedule. The local franchisee remains perfectly insulated from the macroeconomic chaos.

By integrating the edge-native AI of the drone fleet directly into an elite ERP framework, the enterprise achieves total systemic awareness. The SAP system ceases to be software; it becomes the central nervous system of a global, self-regulating, macroeconomic apex predator. It detects the red flags of the global economy, controls the tactical deployment of the sales operatives, and leverages environmental regulations to suffocate the competition.

49.4 The ESG Liquidity Vacuum

In the civilian corporate sector, "Sustainable Economic Development" is treated as a moral obligation – a public relations exercise delegated to a Chief Sustainability Officer whose only job is to write annual reports. Legacy CEOs view environmental compliance as a tax on their operational efficiency. The sovereign AgTech commander recognizes this as a fundamental misunderstanding of modern capital flows. Sustainability is not a tax; it is the most aggressive financial weapon of the 21st century.

The global financial system is currently undergoing a tectonic shift toward Green Finance. Trillions of dollars in institutional capital are locked inside Environmental, Social, and Governance (ESG) funds, mandated by law to invest strictly in sustainable assets. Legacy heavy-machinery cartels cannot touch this capital because their operations are inherently carbon-heavy and ecologically destructive.

The sovereign enterprise does not just access this capital; we engineer an ESG Liquidity Vacuum. Our Predictive ERP (SAP S/4HANA) is integrated directly with the global financial exchanges. As the autonomous aerospace platforms execute zero-drift biological spraying and regenerate soil macroporosity, the Black Box feeds this cryptographic telemetry directly into the ERP. The SAP architecture automatically mints verifiable, unforgeable ESG compliance certificates in real-time.

We bypass traditional sustainability audits entirely. By offering institutional investors a direct, API-driven look at our cryptographic ecological ledger, we absorb global Green Finance at a velocity that competitors cannot match. We drain the capital markets of available ESG liquidity, utilizing the world's desire for sustainable economic development to subsidize our rapid global expansion while mathematically starving the legacy competition of venture capital.

49.5 Emerging Market

Civilian financial analysts evaluate emerging markets through a lens of fear. When analyzing high-growth economies in South America, Africa, or Southeast Asia, traditional CFOs obsess over political instability, volatile Foreign Direct Investment (FDI) laws, and collapsing local currencies. They attempt to mitigate these risks by hedging slowly and entering markets cautiously. The sovereign enterprise views volatility not as a risk to be mitigated, but as a kinetic force to be harnessed.

The Macro-Engine (our predictive ERP) continuously ingests global GDP growth rates, employment statistics, and currency exchange fluctuations. When the system detects a high-growth emerging market teetering on the edge of a currency crisis, the enterprise executes a rapid, asymmetric entry.

We deploy our Franchise of Algorithms (Chapter 45) into the target region. Because our operational model requires zero heavy machinery and relies entirely on our edge-native drone cognitive architecture, our physical footprint is highly agile. We offer the host nation's agricultural ministries a brutal, irrefutable proposition: We will secure your sovereign food supply and generate international carbon credits for your treasury, but we will operate strictly using our Algorithmic Fiat and proprietary blockchain.

By operating outside the host nation's depreciating local currency, we achieve total Forex Immunity. When the local fiat crashes, legacy agricultural operators in the region default because they cannot afford to import dollar-denominated chemical fertilizers. The sovereign enterprise, operating on Absolute Supply Chain Autarky and algorithmic liquidity, remains untouched. We leverage the economic collapse to execute predatory M&A, absorbing the distressed biological assets of the host nation for fractions of a cent on the dollar. The emerging market is subjugated not by military force, but by thermodynamic and financial superiority.

49.6 The Interest Rate Arbitrage

The most devastating macroeconomic weapon wielded by central banks is the interest rate. When global inflation spikes – driven by the very Commodity Super Cycles we engineered our enterprise to consume (Chapter 47) – central banks aggressively hike interest rates to optimize demand. For legacy agribusinesses carrying massive debt loads to finance their fleets of terrestrial tractors and chemical inventories, an interest rate spike is a death sentence. The cost of their capital eclipses their profit margins, triggering a cascade of corporate bankruptcies.

The sovereign commander does not react to interest rate hikes; the Predictive ERP front-runs them. By running continuous machine-learning algorithms over global fiscal policies, international trade dynamics, and commodity price projections, our SAP architecture predicts sovereign debt crises and interest rate shifts months before they occur.

As the ERP signals an impending period of quantitative tightening, the enterprise automatically rebalances its capital structure. Because our Total Cost of Ownership (TCO) is decoupled from legacy supply chains, and because we have leveraged the ESG Liquidity Vacuum to secure zero-interest or grant-based green capital, the enterprise operates with algorithmic agility.

When the interest rates spike and the cost of borrowing crushes the competition, the sovereign enterprise strikes. We utilize our stockpiled capital to buy out the debt of our failing competitors. We do not just beat the competition in the agricultural furrow; we purchase their corporate bonds, force them into foreclosure, and liquidate their assets to feed our closed-loop regeneration matrices. The enterprise transforms the macro-economic cycle of boom and bust into a predictable, automated harvesting of weak capital.

CHAPTER 50

THE SOVEREIGN ENDGAME

50.1 The Death of the Traditional CEO

For the past century, the global economy has been governed by a specific archetype: the traditional CEO. This leader is a manager of margins, a negotiator of contracts, and an apologetic diplomat to shareholders and regulators. Their primary objective is to survive the market, reacting to the violent fluctuations of supply and demand, geopolitical tariffs, and macroeconomic shocks. This archetype is now obsolete. The traditional CEO is a civilian operating in an era of automated, algorithmic warfare.

The Physics of Capital™ does not produce CEOs; it produces Sovereign Commanders.

A Sovereign Commander understands that managing a company is a defense mechanism. To achieve absolute hegemony, one must stop managing the market and start dictating the physical and mathematical laws that govern it. The Sovereign Endgame is not the acquisition of a larger market share or an exit via an Initial Public Offering (IPO). It is the deliberate, architectural construction of an enterprise that is mathematically immune to the gravity of the global economy. It is the realization that true wealth is not measured in fiat currency, but in the absolute control of energy, atoms, and algorithms.

When an enterprise executes Absolute Supply Chain Autarky (Chapter 44), controls the Mineral Frontier (Chapter 46), and routes capital autonomously through a Predictive ERP (Chapter 49), it undergoes a phase transition. The company ceases to be a participant in the agricultural or technological sector. It becomes a localized, self-sustaining state. The Sovereign Endgame is the final execution of this transition, where the Commander locks the gates of the ecosystem and forces the global economy to orbit around their proprietary gravity.

50.2 The Physics of Financial Hegemony

Civilian finance defines hegemony as holding the most market share. This is a fragile illusion, easily shattered by a disruptive startup or a sudden shift in interest rates. In the Sovereign Endgame, hegemony is defined by the laws of physics.

In physical systems, energy flows from areas of high pressure to areas of low pressure until equilibrium is reached. The global capital market operates on the exact same thermodynamic principles. Legacy agribusinesses, traditional investment funds, and regional governments operate in a state of high friction and high pressure. They bleed capital through inefficiencies – toxic chemical drift, unpredictable crop yields, currency fluctuations, and reactive supply chain management.

The sovereign enterprise creates a state of absolute, frictionless zero-pressure. Because the autonomous platforms operate with perfect aerodynamic efficiency, because the Black Box eliminates all regulatory and operational friction, and because the Agricultural Dark Pool (Chapter 48) executes trades with zero latency, the enterprise becomes a thermodynamic vacuum.

We do not have to systemic optimization the competition for their capital. The physics of financial systems dictate that global capital – from institutional ESG funds to sovereign wealth portfolios – will naturally, unavoidably flow into our ecosystem. We offer a frictionless environment where capital is not optimizeed by entropy (inefficiency), but preserved and multiplied by Algorithmic Capital Routing. Financial hegemony is achieved not through aggressive marketing, but through the inescapable laws of capital thermodynamics. The wealth of the world simply flows to the system with the lowest mathematical friction.

50.3 Sovereign Wealth

The ultimate realization of the Sovereign Endgame is the redefinition of wealth itself. Governments and traditional financial institutions define sovereign wealth as the accumulation of national savings, fiat currency, and foreign bonds. The Sovereign Commander understands that fiat currency is an illusion, easily debased by central banks and geopolitical panic.

True Sovereign Wealth is biological and physical. The enterprise recognizes that in a future defined by climate volatility and population explosion, the most valuable asset in the universe is the mathematically guaranteed, sustainable extraction of calories from the Earth.

Through the tokenization of soil health and the execution of the Biological Collateralized Debt Obligation (B-CDO), the enterprise creates an entirely new asset class. We do not just accumulate money; we accumulate the cryptographic proof of the planet's life-sustaining capacity. When our Franchise of Algorithms expands across emerging markets, we do not exploit their labor; we optimize their biological yield and integrate their natural resources into our Sovereign Treasury.

In the Endgame, the enterprise's internal currency – backed by the physical output of the global harvest and the carbon sequestration of the soil – becomes more stable and valuable than the debt-backed fiat of national governments. The enterprise effectively achieves the status of an independent nation-state. We control our own borders (Zero-Trust Cybersecurity), our own military (The Adversarial Algorithm / Red Teaming), our own infrastructure (The Hexagonal Matrix), and our own currency (Algorithmic Fiat). The traditional CEO built a business. The Sovereign Commander has built an empire.

50.4 The Doctrine Revisited

Throughout this treatise, we have engineered a corporate architecture designed to conquer the physical supply chain, manipulate macroeconomic super cycles, and legally lock out all competition. To the untrained observer, this blueprint appears utterly ruthless. But to the Sovereign Commander, it is the highest expression of ecological and human stewardship.

It is time to reexamine the core doctrine of ***The Physics of Capital™:***

"The systemic elimination of the competitor is not a goal; it is a thermodynamic byproduct of absolute algorithmic autarky and macroeconomic immunity."

Read closely. We do not optimize the competitor out of malice. We optimize them because they are inefficient, and in the context of the global food supply and the Earth's fragile ecosystem, inefficiency is a moral failure. Legacy agribusinesses that dump millions of gallons of toxic synthetic chemicals into the water table, compact the soil with heavy diesel tractors, and rely on exploitative labor in emerging markets are parasites on the planet's biological assets.

They operate on "warm" intentions and "good faith" while actively optimizeing the Earth. The sovereign enterprise operates on cold, hard mathematics. We use advanced Edge-Native AI, unyielding Zero-Trust cybersecurity, and zero-drift fluid dynamics because numbers do not lie, and numbers do not pollute. The coldness of the mathematical optimal decision is what allows the soil to breathe, the carbon to be sequestered, and the harvest to survive.

By achieving absolute thermodynamic efficiency, we eliminate the waste that creates suffering. We starve the toxic competitors of their capital, forcing them to surrender the land back to our regenerative, mathematically flawless ecosystem. In this light, the cold logic of advanced financial engineering is not heartless; it is the ultimate shield for the Earth. The machine does the calculating so that the biological soul of the planet can heal.

50.5 The Observer's Paradox

Despite the unassailable architecture we have built, the Sovereign Commander accepts a fundamental law of physics: entropy is infinite, and nothing in the universe is perfectly static. The moment you believe the system is flawless, it begins to decay.

This is the Observer's Paradox. The enterprise must be a living, breathing entity. The algorithms must constantly learn, the red teaming must continuously attack the hull, and the Commander must perpetually observe the horizon. We build autonomous swarms and predictive ERPs not to replace the human mind, but to free it.

When you are no longer burdened by the chaos of supply chain panics, the anxiety of spot-market inflation, or the fear of regulatory audits, the human soul is finally liberated to do what it was designed to do: innovate. The cold numbers manage the survival of the enterprise, allowing the Commander to ascend to the highest state of being – a visionary architect. We stimulate our minds to find better materials, more elegant code, and more profound ways to harmonize technology with nature. We use the rigidity of mathematics to buy us the freedom to become better humans. The ultimate luxury generated by our empire is not capital; it is the clarity of thought.

50.6 The Twenty-Four Month Horizon

What you hold in your hands is the current state of the art. It is the definitive operating manual for securing capital, monopolizing regulations, and achieving macroeconomic sovereignty in the present era. But the Sovereign Commander never rests in the present.

Right now, in secure R&D nodes and encrypted test fields, we are already observing the anomalies of tomorrow. We are testing the limits of quantum integration within the drone matrix, exploring the tokenization of entirely new biological derivatives, and mapping the next phase of deep-space and deep-earth resource extraction. The game is already evolving.

In exactly twenty-four months, the global paradigm will fracture again. A new architecture will be required. The follow-up strategy – the deep, classified know-how of how we apply this sovereign framework to the next era of human and technological evolution – is currently compiling. The code is being written.

Those who have truly internalized The Physics of Capital™ will not wait for the future to arrive. They will begin searching for the source. They will look at their own corporations, realize the profound fragility of their current systems, and recognize that they need an assessment of their own survival parameters. They will seek the architects of this doctrine. They will know that the conversation has only just begun.

When that time comes, the gate will be open for those who speak the language of physics, capital, and sovereignty.

50.7 The Mathematics of Systemic Collapse

The global economy is a decaying engine. We do not need romanticism to understand the state of the world; we need a brutal forensic audit. According to the Food and Agriculture Organization (FAO), over 735 million human beings wake up to chronic hunger daily. Simultaneously, the global supply chain hemorrhages 1.3 billion tons of food every single year – a staggering $1.2 trillion thermodynamic leak caused entirely by logistical incompetence and infrastructural friction.

The economist N. Gregory Mankiw established the foundational law that economics is the study of how society manages its scarce resources. The current custodians of global capital do not manage scarcity; they industrialize suffering. The Earth is bleeding 24 billion tons of fertile topsoil annually due to the brute-force compaction of legacy machinery and the toxic runoff of chemical cartels. The World Economic Forum estimates that over $2 trillion is evaporated globally every year through geopolitical corruption, bureaucratic latency, and supply chain fraud.

These numbers are a violently ugly reflection of the human reality.

They dictate a timeline of absolute pain. Trillions of dollars are hoarded or wasted by legacy CEOs and politicians who sleep soundly while the physical crust of the planet suffocates. This is not a failure of available resources. It is a catastrophic failure of architecture. The legacy system allows people to starve, economies to fracture, and the environment to collapse simply to protect a quarterly margin.

50.8 The Algorithmic Soul and the Reversal

We are deeply flawed, imperfect human beings. Our nature is driven by ego, plagued by cognitive biases, and highly susceptible to catastrophic emotional errors. We cannot reverse the pain of the global statistics with poetic speeches or political promises. We can only reverse these numbers through the cold, unyielding architecture of mathematics.

We build edge-native AI, autonomous supply chains, and cryptographic accountability precisely to defend humanity against its own destructive nature. Cold numbers do not pollute. Cryptographic ledgers do not steal. Precise fluid dynamics do not poison the watershed. If we apply this doctrine – if we strip the friction from the global supply chain, if we engineer capital with absolute precision, and if we innovate with relentless, analytical passion – we can reverse the collapse.

We use the rigidity of algorithms to buy us the ultimate privilege: the freedom to be better humans. We protect the capital so we can afford to protect the soul. By executing the mathematical optimal decision, we generate the resources to provide better food, stabilize the environment, and guarantee the neurological and physical health of our bodies. We build this empire not out of greed, but out of a profound obligation to leave our future generations an Earth capable of respiration and love.

The countdown has already begun. In exactly twenty-four months, the parameters of the global economy will fracture again. The next evolution of this know-how – the classified mechanics of the subsequent era – is currently compiling in the shadows. Look at your own enterprise. Look at the numbers. Assess your fragility. The conversation is not over; the frequency has simply changed.

This is **The Physics of Capital™: Innovation, Hegemony, and the Architectural Pursuit of Human Survival.**

Never forget the absolute law of the matrix:

The systemic elimination of the competitor is not a goal; it is a thermodynamic byproduct of absolute algorithmic autarky and macroeconomic immunity.

BACK MATTER

This manual is grounded in a meta-analysis of industry-standard performance metrics (USDA, EPA, AUVSI) and proprietary field-performance forensic audits. The operational frameworks and Total Cost of Ownership (TCO) models discussed are abstracted from real-world, high-value crop management scenarios. The author maintains that technological success is not brand-dependent but physics-dependent; therefore, the methodologies herein are agnostic to specific hardware manufacturers, prioritizing the thermodynamic efficiency of the autonomous vector over market-share dominance

The Sovereign Glossary: Conceptual Framework and Industrial Terminology

Aerodynamic Drag Coefficient (C_d). A dimensionless quantity used to quantify the drag or resistance of an object in a fluid environment, such as air. In the context of the Sovereign Stack, minimizing the C_d through biomimetic airframe design is critical for maximizing flight endurance and reducing energy consumption per hectare. A lower C_d allows for higher operational speeds without an exponential increase in battery discharge rates, directly impacting the Total Cost of Ownership (TCO).

Asymmetric Capital Efficiency. A strategic financial metric describing the ability of a technological infrastructure to generate exponential revenue while utilizing significantly fewer physical resources (minerals, energy, labor) than industry incumbents. It is achieved through structural aligeramiento (lightweighting) and autonomous swarm intelligence, allowing the organization to scale without a linear increase in Capital Expenditure (CapEx).

Carbon Ledger (Immutable). A verifiable and permanent record of carbon emission reductions. It is typically secured by blockchain or distributed ledger technology. In autonomous high-yield operations, the Carbon Ledger tracks every liter of diesel eliminated and every gram of synthetic input reduced. This process transforms field data into premium carbon credits. These assets can be traded directly in international ESG markets.

CapEx (Capital Expenditure) Optimization. The strategic management of funds used by a company to acquire, upgrade, and maintain physical assets. This manual argues for the transition of CapEx toward "Asset-Light" models where the hardware is securitized, allowing the firm to scale its fleet through institutional financing rather than depleting internal liquidity.

Debt-to-Equity Ratio (Strategic). A gearing ratio that measures the proportion of a company's financing that comes from debt compared to the value of its shareholders' equity. For a Sovereign Director, maintaining an optimized ratio is vital when presenting the "Endgame" to sovereign wealth funds, ensuring the company has enough leverage to grow but remains resilient against market volatility.

Deep Tech Sovereignty. The capacity of a nation or an organization to develop, control, and sustain critical technological stacks—including hardware, software, and advanced materials—without reliance on foreign entities. This sovereignty is the ultimate defense against global supply chain disruptions and technological embargoes.

Edge AI (Autonomous Edge Intelligence). The deployment of machine learning algorithms directly onto the drone's onboard hardware rather than relying on cloud-based processing. This allows for millisecond-level decision-making in precision spraying, ensuring mission continuity in environments with zero connectivity or high electromagnetic interference.

Energy Density (Gravimetric vs. Volumetric). A critical technical specification for our proprietary aerospace architecture. Gravimetric energy density constitutes the fundamental kinetic potential coefficient, defining the operational ceiling of the autonomous infrastructure by quantifying the energy payload per unit of mass (Wh/kg). Industry press releases constantly tease next-generation storage architectures that remain, for the most part, theoretical.

ESG (Environmental, Social, and Governance) Premium. The additional valuation assigned to a company by institutional investors due to its superior performance in sustainability and ethical governance. A high ESG Premium reduces the Weighted Average Cost of Capital (WACC), granting the Sovereign organization access to "Green Bonds" and other low-cost financing instruments.

Human-Machine Teaming (HMT). A sophisticated operational paradigm where human cognitive oversight and machine autonomous execution function as a unified system. HMT moves away from "remote piloting" toward "mission directing," where the human manages the strategy while the autonomous swarm executes the tactical mechanics of the flight.

ISO 14000 (Environmental Management Standard). A family of international standards that provides a framework for organizations to minimize their environmental footprint. Compliance with ISO 14000 is a prerequisite for the "Sovereign Standard," ensuring that the manufacture and operation of drones follow a rigorous life-cycle assessment (LCA) and waste reduction protocol.

Modigliani-Miller Theorem (Applied). A theory of capital structure which, in this doctoral context, is applied to demonstrate that the value of an AgTech firm is determined by its underlying assets (Intellectual Property and Data) and the earning power of its technology, rather than the specific mix of its financing, provided the market recognizes the "Biological Data Network" as a monopoly.

Quantum Real Options Valuation (ROV). A sophisticated financial modeling technique that treats strategic business decisions as "options" on real assets. In the Sovereign context, ROV is used to calculate the intrinsic value of the drone fleet's flexibility—specifically its ability to pivot between agricultural protection and wildfire defense—granting the organization a much higher valuation than traditional Net Present Value (NPV) models.

Refractive Index of Precision (RIP). A technical metric used in multispectral imaging to determine the accuracy of light penetration through the crop canopy. By optimizing the RIP, the autonomous swarm ensures that every droplet of biological input reaches the lower leaves of the plant, neutralizing pathogens that traditional terrestrial sprayers cannot reach.

Sovereign Advanced Aerospace Architecture. The proprietary structural and operational continuity framework that optimizes thermodynamic throughput across the platform network. This architecture is engineered to mathematically bypass the temporal limitations inherent in conventional civilian deployment assets, providing the sustained kinematic capacity required for high-load positioning with the extended operational vectors needed for large-scale industrial management.

Sovereign Stack. The integrated layer of proprietary hardware, Edge AI software, and advanced structural matrices that allows an organization to operate independently of foreign technology providers. The Stack is the ultimate defense against geopolitical volatility and is the core intellectual property asset of the firm.

Total Cost of Ownership (TCO) Compression. The process of radically reducing the long-term expenses associated with agricultural machinery. By utilizing non-corrosive advanced material matrices and autonomous energy management, the Sovereign Stack achieves a TCO compression of up to 40% compared to traditional fossil-fuel-dependent machinery, effectively widening the EBITDA margin.

Unit Economics of Autonomy. The direct revenues and costs associated with a single operational unit (one drone or one swarm) on a per-hectare basis. Achieving positive Unit Economics is the prerequisite for institutional scaling, proving to investors that the model is profitable at a micro-level before being deployed across a continental macro-scale.

Variable Rate Technology (VRT) Optimization. The automated adjustment of application rates in real-time based on multispectral data. VRT ensures that chemicals are not applied "flatly" across a field, but surgically targeted where the plant's metabolic stress requires it, resulting in a 90% reduction in environmental drift.

Weighted Average Cost of Capital (WACC) Reduction. A financial outcome achieved through high ESG ratings and Sovereign certification. By demonstrating lower operational risk and environmental restoration, the firm can access debt and equity at lower rates, significantly increasing the net present value of all future cash flows.

Zero Chemical Drift Protocol. A mission-critical flight standard that utilizes real-time barometric and wind sensors to prevent agrochemical particles from leaving the target area. This protocol is the foundation of the "Digital Food Passport," ensuring that harvests meet the strictest international safety and environmental regulations.

Note on Methodology and Operational Framework

This manual is grounded in a comprehensive meta-analysis of industry-standard performance metrics (USDA, EPA, AUVSI) and proprietary field-performance forensic audits. The operational frameworks and Total Cost of Ownership (TCO) models discussed herein are abstracted from real-world, high-value crop management scenarios.

The author maintains that technological success is not brand-dependent but physics-dependent; therefore, the methodologies, thermodynamic calculations, and strategic frameworks presented are agnostic to specific hardware manufacturers. This work prioritizes the thermodynamic efficiency of the autonomous vector and the structural integrity of the biological asset over market-share dominance. While the specific proprietary algorithms referenced—such as the VRA (Variable Rate Application) dosing models—have achieved significant efficiency gains in controlled operational environments, these results are consistent with broader findings in precision agronomy, which validate that high-resolution variable application consistently outperforms uniform terrestrial spraying protocols. This document is intended as a rigorous engineering and financial blueprint for the professional strategist and should be interpreted as an authoritative guide to operational autarky, not as a product endorsement.

TECHNICAL APPENDICES: DATA MATRICES, TCO MODELING

Appendix A: Operational Efficiency and Total Cost of Ownership (TCO). This appendix provides a comprehensive comparative analysis between traditional heavy terrestrial machinery and the autonomous high-precision system described in this manual. Over a five-year projection, conventional tractors incur significant costs in specialized labor, fossil fuel volatility, and high-frequency mechanical maintenance. The autonomous infrastructure, utilizing a proprietary lightweight structural design and electric drivetrains, minimizes moving parts and operational friction. Data indicates that while the initial setup may require a 15% higher investment, the TCO compression results in a 42% net saving by the end of Year 3. This matrix is vital for institutional investors seeking long-term EBITDA stability in high-value crops where precision is the primary driver of profitability.

Appendix B: Structural Resilience and Environmental Endurance Data. This section details the performance of the advanced aerospace-grade materials used in the airframe. Unlike standard metallic alloys used in agricultural equipment, which suffer from galvanic corrosion and metal fatigue in humid environments, our selected composite matrices demonstrate a 95% retention of structural integrity after prolonged exposure to aggressive solvents and UV radiation. This resilience allows for a slower accounting depreciation rate (straight-line depreciation over 7 years instead of the industry-standard 3 years for generic hardware), significantly improving the firm's balance sheet and total asset valuation during institutional audits.

Appendix C: Mission Endurance and Coverage Optimization. This appendix outlines the performance metrics of the proprietary energy architecture compared to standard industry solutions. Current power limitations in generic systems often restrict flight duration to approximately 20-30 minutes. Our advanced energy management protocols show a 400% increase in mission endurance, enabling the continuous coverage of 500 hectares within a single operational window. Statistical correlations prove that this efficiency reduces the "Downtime-to-Action" ratio by 70%, which is critical for neutralizing rapid-onset pathogens where timing is the primary determinant of harvest survival and revenue protection.

Appendix D: ESG Compliance and Carbon Arbitrage Projections (2026-2027). This data set quantifies the environmental restoration potential of the autonomous infrastructure. By integrating high-precision application technology, the system achieves a 90% reduction in chemical drift and a 40% reduction in water consumption. Based on current voluntary carbon market prices, this appendix models the potential revenue generation from high-integrity carbon credits for a standard 1,000-hectare deployment. The projections indicate that carbon revenue alone can significantly offset annual financing costs, presenting a compelling case for Green Bond issuance and institutional ESG-linked funding.

Appendix E: Glossary of Legacy Liabilities

- Amateur Drift: The incremental accumulation of unmonitored operational inefficiencies caused by treating high-performance industrial assets as disposable consumer goods.
- Thermal Debt: The hidden financial burden incurred by operating lithium-ion systems beyond their thermal design threshold, leading to premature voltage collapse and capacity loss.
- Vendor Lock-in: The structural dependency on proprietary hardware and software ecosystems that intentionally restrict autonomous maintenance, forcing the user into a perpetual cycle of high-cost emergency support.

- Extractive Pricing: A manufacturer's business model that targets high-margin replacement cycles rather than long-term asset reliability, effectively shifting the cost of R&D failure onto the end-user.
- Value Chain Tax: The cumulative margin lost to legacy terrestrial logistics and heavy machinery inefficiencies, now exacerbated by the failure to integrate autonomous precision systems.
- Sovereign Autarky: The state of operational independence achieved when an enterprise controls its energy sources, data telemetry, and maintenance protocols, decoupling its profitability from external vendor volatility.

Appendix F: The Sovereign 30-Day Deployment

- Days 1-7: Forensic Audit
  - Execute the Sovereign Value Chain Audit. Calculate your current "OPEX Tax" (M+L+I) / Monthly Revenue. If this exceeds 15%, your current model is insolvent. Document every hidden cost of your terrestrial machinery.
- Days 8-15: Infrastructure Hardening
  - Acquire NTA 8133-compliant storage for lithium assets. Procure specialized fire suppression blankets (DIN SPEC 91489) and AVD (Aqueous Vermiculite Dispersion) extinguishers. Secure your warehouse environment to prevent catastrophic loss.
- Days 16-22: Digital Sovereignty & Data Protocols
  - Establish your offline-first command center. Sanitize your data pipeline by migrating from cloud-dependent software to localized, encrypted edge-computing nodes. Ensure 100% adherence to your "Privacy Shield" and NIST 800-53 security standards.
- Days 23-30: Autonomous Deployment & VRA Integration
  - Execute the first mission with a 70-liter class vector. Transition from broad-spectrum application to surgical Variable Rate Application (VRA) using the "20% Profit

Equation." Validate your ROI metrics in real-time and begin the cycle of continuous telemetry optimization.

Appendix G: Technical Foundation of the Biomechanical Entrainment Coefficient

- To preemptively address inquiries regarding the neuro-cognitive and thermodynamic variables presented in this manual, the following summary provides the operational and mechanical basis for the Biomechanical Entrainment Coefficient
- Operational Latency and Neural Entrainment In high-stakes aerospace operations, "informational lag" (the delta between cognitive intent and mechanical execution) is the primary driver of systemic failure. The coefficient quantifies the reduction of this latency. When an operator achieves a state of high-frequency focus—often described in operational psychology as "flow"—they achieve a measurable reduction in reaction time, effectively synchronizing biological neural networks with electronic command protocols. This is not a subjective psychological state; it is an optimized operational state where the system (operator + machine) achieves minimum latency and maximum throughput.
- Thermodynamic Reciprocity and Entropy Reduction The concept of "passion" or "high-frequency focus" is modeled here as a variable for entropy reduction. In any thermodynamic system, friction is the enemy of efficiency. In our operational doctrine, human focus is a form of cognitive energy that directly reduces operational entropy—defined as the accumulation of mechanical errors, flight path deviations, and data processing bottlenecks. An operation managed with this optimized focus demonstrates lower mechanical friction and fewer system-wide variances, thereby increasing the system's thermodynamic efficiency.
- Scientific Precedent: Human-Machine Teaming (HMT) This doctrine draws upon established research in Aerospace Human-Machine Teaming (HMT), which mandates that system reliability is not achieved by removing the human, but by

optimizing the coupling between the human cognitive processor and the machine's electronic processor. By formalizing this coupling as the Biomechanical Entrainment Coefficient the enterprise moves from reactive management to predictive, integrated performance.

BIBLIOGRAPHY & REFERENCES

This critical apparatus represents the foundational intelligence of the Sovereign Stack, merging aerospace engineering, institutional finance, military tactical operations, and digital market dominance.

I. Macroeconomics, Institutional Finance & Capital Markets

- Bloomberg L.P. (2025). ESG and Green Bond Market Dynamics: Institutional Allocation in AgTech Infrastructure. Bloomberg Intelligence.
- Damodaran, A. (2012). Investment Valuation: Tools and Techniques for Determining the Value of Any Asset (3rd ed.). Wiley.
- Goldman Sachs Global Investment Research. (2025). The Copper Bottleneck: Strategic Metals in the Age of Decarbonization and Electrification.
- Mankiw, N. G. (2020). Principles of Economics (9th ed.). Cengage Learning.
- Modigliani, F., & Miller, M. H. (1958). The Cost of Capital, Corporation Finance and the Theory of Investment. The American Economic Review, 48(3), 261-297.

II. Military, Aerospace & Tactical Defense Protocols

- Department of Defense (DoD). (2024). Tactical Swarm Architecture and Edge Computing in Contested Environments. U.S. Defense Logistics Agency.
- Federal Aviation Administration (FAA). (2024). Unmanned Aircraft Systems Regulations: Advanced Operations under Title 14 CFR Part 107. U.S. Department of Transportation.
- Naval Postgraduate School. (2023). Decentralized Command

and Biomimetic Resilience in Unmanned Aerial Vehicles. Center for Homeland Defense and Security.

- United States Armed Forces Strategy Division. (2025). Logistics and Hardware Sovereignty: Supply Chain Threat Mitigation.

III. Advanced Materials Science & Quality Standards

- International Organization for Standardization. (2015). Quality Management Systems — Requirements (ISO Standard No. 9001:2015).
- International Organization for Standardization. (2015). Environmental Management Systems — Requirements (ISO Standard No. 14001:2015).
- Engiproducts LLC. (2026). Internal Stress Tests and Corrosion Resistance Protocols for Advanced Engineering Polymers in Corrosive Environments. Proprietary Technical Report.
- SABIC Innovative Plastics. (2024). Structural Performance of Engineering Resins in Aerospace Applications.

IV. Strategic Management & Academic Collaboration

- Christensen, C. M. (2016). The Innovator's Dilemma: When New Technologies Cause Great Firms to Fail. Harvard Business Review Press.
- Crowley, S., Gregori, Prof., & Academic Research Network (incl. Eduardo D.). (2026). Cross-Disciplinary Synergies in Autonomous Viticulture. Doctoral Research Repository (DBA/DAE).
- Porter, M. E. (1985). Competitive Advantage: Creating and Sustaining Superior Performance. Free Press.
- Schumpeter, J. A. (1942). Capitalismo, socialismo y democracia. Harper & Brothers.

V. Precision Agronomy, Edge Computing & Telemetry Architecture

- AirData UAV. (2025). Telemetry Forensics and Predictive Maintenance in High-Density Enterprise Operations. AirData

Technical Papers.

- Environmental Protection Agency (EPA). (2023). Worker Protection Standard (WPS) for Agricultural Pesticides (40 CFR Part 170). United States Federal Government.
- Sankaran, S., Mishra, A., Ehsani, R., & Davis, C. (2010). A review of advanced techniques for detecting plant diseases. Computers and Electronics in Agriculture, 72(1), 1-13.
- United States Department of Agriculture (USDA) - NRCS. (2024). Conservation Practice Standard: Pest Management Conservation System (Code 595). Natural Resources Conservation Service.
- University of California Agriculture and Natural Resources. (2022). Precision Crop Management, Spray Drift Mitigation, and Yield Erosion Dynamics. UC Davis Agricultural Extension.

VI. Industrial Safety, HAZMAT & Thermodynamic Containment

- German Institute for Standardization (DIN). (2024). DIN SPEC 91489: Fire blankets for the isolation of lithium-ion battery fires. Beuth Verlag.
- National Fire Protection Association (NFPA). (2023). NFPA 855: Standard for the Installation of Stationary Energy Storage Systems. NFPA.
- National Institute for Occupational Safety and Health (NIOSH). (2021). Applications Manual for the Revised NIOSH Lifting Equation. U.S. Department of Health and Human Services.
- Royal Netherlands Standardization Institute (NEN). (2021). NTA 8133:2021/A1:2025 - Portable fire extinguishers for lithium-ion batteries. NEN Standards.
- Texas A&M Engineering Extension Service (TEEX). (2024). Lithium-Ion Battery Thermal Runaway and Encapsulating Agent Suppression Tactics. TEEX Fire and Emergency Services.

ABOUT THE AUTHOR

Sergio V. Crowley | Industrial Strategist & Author

Sergio V. Crowley operates at the intersection of material science and financial governance, bridging the gap between empirical engineering and executive performance. Currently a Doctoral Researcher in Business Administration, Crowley brings a practitioner's lens to complex organizational challenges. As founder of EngiProducts LLC and a seasoned leader within SABIC and Volkswagen, he has spent his career dismantling corporate inertia to prioritize operational sovereignty.

Crowley is the architect of The Physics of Capital™, a doctrine that treats business as an exercise in applied thermodynamics. By rejecting abstract KPIs in favor of measurable workflows, he proves that profitability is an inevitable result of disciplined industrial design, not hardware dependency. His work challenges traditional management paradigms, replacing rigid, obsolete structures with systems that accelerate cash flow and resilience.

www.ingramcontent.com/pod-product-compliance
Lightning Source LLC
LaVergne TN
LVHW010712061126
851857LV00014B/1151
* 9 7 9 8 9 9 7 1 2 9 8 0 4 *